Jürgen Albiez

Finite element simulation of dislocation based plasticity and diffusion in multiphase materials at high temperature

Schriftenreihe
Kontinuumsmechanik im Maschinenbau
Band 14

Karlsruher Institut für Technologie (KIT)
Institut für Technische Mechanik
Bereich Kontinuumsmechanik

Hrsg. Prof. Dr.-Ing. habil. Thomas Böhlke

Eine Übersicht aller bisher in dieser Schriftenreihe erschienenen Bände finden Sie am Ende des Buchs.

Finite element simulation of dislocation based plasticity and diffusion in multiphase materials at high temperature

by
Jürgen Albiez

KIT Scientific Publishing

Dissertation, Karlsruher Institut für Technologie
KIT-Fakultät für Maschinenbau

Tag der mündlichen Prüfung: 31. Januar 2019
Referenten: Prof. Dr.-Ing. Thomas Böhlke
Prof. Dr.-Ing. Martin Heilmaier

Impressum

KIT Scientific Publishing

Karlsruher Institut für Technologie (KIT)
KIT Scientific Publishing
Straße am Forum 2
D-76131 Karlsruhe

KIT Scientific Publishing is a registered trademark
of Karlsruhe Institute of Technology.
Reprint using the book cover is not allowed.

www.ksp.kit.edu

Print on Demand 2019 – Gedruckt auf FSC-zertifiziertem Papier

ISSN 2192-693X
ISBN 978-3-7315-0918-9
DOI 10.5445/KSP/1000092297

Finite element simulation of dislocation based plasticity and diffusion in multiphase materials at high temperature

Zur Erlangung des akademischen Grades

Doktor der Ingenieurwissenschaften

der KIT-Fakultät für Maschinenbau

Karlsruher Institut für Technologie (KIT)

genehmigte

Dissertation

von

Dipl.-Ing. Jürgen Albiez

Tag der mündlichen Prüfung:	31.01.2019
Hauptreferent:	Prof. Dr.-Ing. Thomas Böhlke
Korreferent:	Prof. Dr.-Ing. Martin Heilmaier

Zusammenfassung

Die gerichtete Erstarrung von eutektischen Legierungen ermöglicht die Herstellung vielversprechender *in situ* Verbundmaterialien für zukünftige Hochtemperaturbauteile, speziell für die Verwendung unter mechanischer Belastung. Um ein umfassendes Verständnis dieser Legierungen zu erhalten und weitere Optimierungsmöglichkeiten aufzudecken, wird ein Simulationsmodell basierend auf der Einkristall-Plastizitäts-Theorie eingeführt. Zur physikalischen Darstellung des Übergangs von der theoretischen Festigkeit zur bekannten Taylor-Festigkeit wird ein neues Verfestigungsgesetz eingeführt. Dieses kann unter anderem zur Beschreibung der *in situ* Verstärkungsphase von gerichtet erstarrtem NiAl-9Mo und NiAl-31Cr-3Mo verwendet werden. Das Kriechverhalten der gerichtet erstarrten NiAl-9Mo Legierung kann mit diesem Verfestigungsgesetz und einem Kristallplastizitätsmodell sowohl für (i) unterschiedliche Belastungen als auch für (ii) unterschiedliche Temperaturen und für (iii) verschiedene Faserdurchmesser sehr gut dargestellt werden. Das Kristallplastizitätsmodell wird anschließend zu einem Gradienten-Plastizitäts-Modell erweitert. Auf diese Weise können die Effekte der Faser-Matrix Grenzfläche und der geometrisch notwendigen Versetzungen in der Simulation berücksichtigt werden. Die „interface-enriched generalized finite element method" wurde hierfür im Kontext der Gradienten-Plastizität angewendet. Eine spannungsabhängige Grenzflächenfließbedingung ist notwendig, um den Übergang eines rein elastischen Verhaltens zu einem elastisch-plastischen Verhalten der Fasern darstellen zu können. Es zeigt sich, dass das mechanische Verhalten der Faser und der Faser-Matrix Grenzfläche das Kriechverhalten des gesamten Verbundes wesentlich beeinflussen. Darüber hinaus wird die Änderung der Mikrostruktur infolge von Diffusion untersucht. Hierfür wird ein Simulationsmodell eingeführt, welches die mechanischen Gleichgewichtsbedingungen und die Cahn-Hilliard Diffusionsgleichung miteinander koppelt. Die Simulationsergebnisse zeigen, dass die Orientierung der Lamellen in Bezug zur Belastungsrichtung eine Rolle für den Diffusionsfluss spielen.

Abstract

Directionally solidified (DS) NiAl based eutectic alloys are promising candidates for further structural applications at elevated temperatures. A single-crystal plasticity model is established for the understanding and the optimization of DS NiAl-9Mo and NiAl-31Cr-3Mo eutectics. To consider the transition from theoretical to bulk strength, a hardening model was introduced and used to describe the strength of the reinforcing phases. The creep behavior of the DS NiAl-9Mo composite could be well described by the single-crystal plasticity model, even for (i) a change of the applied stress, (ii) a change of the temperature, and (iii) a change of the fiber diameter. To include the effects of the fiber-matrix interface and the geometrically necessary dislocations within the simulations, the single-crystal plasticity model was extended to a gradient crystal plasticity model. It was realized by applying the interface-enriched generalized finite element method, which was originally used to describe discontinuities resulting from cracks. Further, to rationalize the transfer from a purely elastic to an elasto-plastic behavior of the reinforcing fibers, the stress dependent interface flow rule was crucial. It was found that the fiber phase as well as the fiber-matrix interface play a particularly important role on the creep behavior of the composite. Moreover, to account for microstructural changes due to material flux, a coupled diffusional-mechanical simulation model was introduced. Here, the focus lay on the investigation of the influences of an inhomogeneous strain distribution on the coarsening process, thus, the Cahn-Hilliard diffusion equation was coupled with the balance equation of linear momentum. It was shown that the lamellar orientation with respect to the external load can influence the coarsening velocity.

Acknowledgments

I would like to express my gratitude to Prof. Thomas Böhlke for offering me a position as an academic employee and for supporting me during the past years. Furthermore, I would like to thank Prof. Martin Heilmaier for co-advising this work, for his support, and for the discussions within the collaborative work. I would also like to thank Dr. Daniel Weygand for the fruitful discussions regarding the joint publications.
Moreover, I thank all members of the IMD research group (VH-KO-610), especially Ioannis Sprenger for providing experimental data. In addition, I thank Anke Schmitt for the collaboration and the discussions. Furthermore, I thank Katrin Längle for her excellent work and her assistance.
I also would like to thank all my colleagues at the ITM for the very good company during working time as well as during free time. Personally, I thank my wife for her encouragement and support.
The support of the Helmholtz School Integrated Material Development for Novel High Temperature Alloy at the Karlsruhe Institute of Technology (KIT), funded by the Initiative and Networking Fund of the Helmholtz Association (VH-KO-610), is gratefully acknowledged.
Finally, I would like to acknowledge the support by the German Research Foundation (BO 1466/12-2) as well as the support by the state of Baden-Württemberg through bwHPC.

Karlsruhe, March 2019 Jürgen Albiez

Contents

Chapter 1

Introduction

1.1 Motivation and objective of the thesis

The development of new high temperature materials for structural applications such as, e.g., turbine blades is a challenging task in material science and engineering. However, this challenge is worthwhile as the potential impact of these materials on turbine engines is highly relevant from an economical and an ecological point of view. Over the past years, the improvement of these engines was mainly driven by using thermal-barrier coatings and elaborating improved cooling systems. Nevertheless, the deviation from the ideal performance further degrades, particularly due to the necessary effort of cooling. This also results from the fact that the operating temperature of engines is further approaching the melting temperature of the constituent materials. Consequently, the discovery of a new material with a higher melting temperature is indispensable, to close the gap and to enhance the performance. Here, numerical analysis and simulations can essentially support and abbreviate the development process. In addition, they deliver an insight into the material's behavior, which helps to understand the high temperature process in general. This gained knowledge can then be used for future developments and optimizations.
Consequently, the movitation for this work is based on the development of a new high temperature material for structural applications. It was

embedded in the *Helmholtz School Integrated Materials Development for Novel High Temperatures Alloys* at the *Karlsruhe Institute of Technology (KIT)*. The research focus was set on NiAl-based directionally solidified eutectics, where the NiAl matrix is reinforced by a refractory metal, e.g., Cr or Mo. These composites reveal good high temperature properties in combination with a higher melting temperature compared to the currently used materials and, thus, are of great interest for further high temperature structural applications. Even though these directionally solidified composites have been studied at room temperature and also at high temperatures, a physically motivated model which rationalizes the composites' behavior is still lacking. The main objective of the presented work is, therefore, to establish and verify physical models which are applicable for NiAl-based directionally solidified eutectics with lamellar or fibrous *in situ* reinforcements. These models should describe the thermo-mechanical behavior under various thermal and mechanical loads and for different withdrawal rates during directional solidification. Further, they should provide an insight into the composites' behavior and reveal the deformation-controlling phases and mechanisms. Preferably, the simulation models are able to clarify and to explain uncertainties within experimental observations. Moreover, they can be used to optimize and to further enhance the composites' high temperature properties. An additional purpose of this work is, finally, to reveal the effects from external loads on the diffusion process leading to microstructural changes.

Within this work, several originalities were developed, whereas the main highlights are:

- The motivation of a size-dependent transition law between different strengthening mechanisms. Compared to the preexisting transition laws, it fulfills more experimental observations and is valid for an extended range of dislocation densities. It describes the strength, decreasing from the theoretical strength for an essentially vanishing dislocation density and reaching the normal square root Taylor hardening for high densities. In addition, it rationalizes several experimental findings on the material strength. Finally, it was used to describe the reinforcing phases of the DS eutectics, however, it is also valid for other materials.

- The development of a single-crystal plasticity model to describe the creep behavior of NiAl-based directionally solidified eutectics with *in situ* reinforcements, validated for NiAl-9Mo and NiAl-31Cr-3Mo. The model is applicable for various thermal and mechanical loads and for different withdrawal rates during directional solidification. The simulation model allows an insight into the materials' behavior and reveals the creep-controlling phase.
- The extension of the local simulation model by the gradient crystal plasticity model to account for the effects of geometrically necessary dislocations. The model incorporates the pile-up stress at the fiber-matrix interface, which leads to the transfer/activation of slip. The corresponding interface flow rule is stress dependent and was validated by comparing simulated creep curves with experimentally measured ones from literature.
- The application of the interface-enriched generalized finite element method (IGFEM) in the context of gradient plasticity. This method allows for discontinuities at interfaces and is well-suited for moving interfaces. It was validated in the context of gradient plasticity by comparing the simulation results with an analytical solution.
- The preliminary results of the Cahn-Hilliard diffusion coupled with the balance equation of linear momentum. The coarsening process of colonies, which contain fine lamellas, is numerically investigated and the impact of an external load on the diffusion process is discussed.

1.2 Outline of the thesis

The experimental background of the directionally solidified eutectics is given in Chapter 2, together with an overview of the characterization of the materials' properties at elevated temperatures. Fundamental informations about the plastic deformation based on dislocations and the basics of continuum mechanics are presented in Chapter 3 and Chapter 4, respectively. The numerical implementations are described in Chapter 5 and are necessary for the realization of the simulation models. In Chapter 6, the transition law of the material strength between different strengthening mechanisms is motivated. This transition law is crucial for the single-crystal plasticity model to describe the creep behavior of DS NiAl-based eutectics presented in Chapter 7. An extension of this local model is given in Chapter 8 by a gradient crystal plasticity model. Amongst others, it considers the effects of the geometrically necessary dislocations and the slip transfer/activation at the fiber-matrix interface. To account for a microstructural change due to diffusion, the coupled diffusional-mechanical simulation model is introduced in Chapter 9. Here, the focus lies on the investigation of the influences of an inhomogeneous strain distribution on the coarsening process. Finally, the work is concluded and the main highlights are summarized in Chapter 10. Further, it includes also a brief discussion of remaining open questions and possible further investigations.

1.3 Notation

In this work, a direct tensor notation is preferred. Vectors and second-order tensors are denoted by lower- and uppercase bold letters, e.g., $\boldsymbol{a}$ and $\boldsymbol{A}$, respectively. The scalar product is marked by $\boldsymbol{A} \cdot \boldsymbol{B} = \mathrm{sp}(\boldsymbol{A}^{\mathsf{T}}\boldsymbol{B})$, whereas the dyadic product is marked by $\boldsymbol{A} \otimes \boldsymbol{B}$. The linear mapping of a first-order tensor by a second-order tensor and the linear mapping of a second-order tensor by a fourth-order tensor $\mathbb{C}$ is identified as $\boldsymbol{t} = \boldsymbol{\sigma n}$ and $\boldsymbol{A} = \mathbb{C}[\boldsymbol{B}]$, respectively. A superimposed dot, e.g., $\dot{\check{\gamma}}$, indicates the material time derivative of the quantity. $\langle \cdot \rangle$ stands for the Macaulay bracket with the properity $\langle \cdot \rangle = \max(\cdot, 0)$. The jump of a quantity is denoted by $[\![\check{\gamma}]\!] = \check{\gamma}^{+} - \check{\gamma}^{-}$ and the normal vector of the interface $\boldsymbol{n}_{\Gamma}$ is pointing from $\mathcal{V}^{-}$ to $\mathcal{V}^{+}$, whereas the mean value of a quantity is denoted by $\langle\!\langle \check{\phi} \rangle\!\rangle$. The Frobenius norm is identified as $||\boldsymbol{A}|| = \sqrt{\boldsymbol{A} \cdot \boldsymbol{A}}$.

Chapter 2

Experimental background of directionally solidified eutectics for high temperature applications

2.1 Characterization of creep behavior

2.1.1 General overview

Under an applied stress, a material deforms permanently at elevated temperatures, even though the applied stress is lower than the yield point of this material. This phenomenon is called creep and usually appears for $T > 0.4\,T_{\mathrm{m}}$, where T stands for the temperature and T_{m} for the melting temperature of the specific material. The most common way to examine a material's creep behavior is the uni-axial creep test. Here, a specimen is heated up to a specific temperature and loaded by a normal stress $\bar{\sigma}$. Temperature and applied stress are kept constant during the creep test and the strain ε in loading direction is plotted versus time t. An exemplary creep curve is shown schematically in Fig. 2.1 a). After an instantaneous elastic strain as a response to the applied stress, the material starts to deform plastically. The creep curve can be divided into three typical creep regimes: the primary I, the secondary II, and the tertiary III creep regime. In the primary creep regime, the creep rate decreases to the minimum creep rate. The secondary creep regime is characterized by an approximately constant creep rate, which increases in the tertiary creep regime. Finally, the creep curve is limited by the creep rupture of the specimen. Another possibility to depict the creep curve is shown in Fig. 2.1 b), where the creep rate is plotted over the creep strain. The creep strain is defined as the difference between the measured strain and the elastic strain (Naumenko and Altenbach, 2016). The classical creep curve (marked with Ⓐ) contains all three creep regimes. Such creep curves can be observed for several materials, e.g., for pure metals (Naumenko and Altenbach, 2016). The creep rate in the secondary creep regime stays approximately constant over a wide range of creep strain and is called steady-state creep rate $\dot{\varepsilon}_{\mathrm{ss}}$. In this regime, the hardening and the softening/recovery effects are canceled out by each other. For many structural materials, however, such a secondary creep regime is not visible (Naumenko and Altenbach, 2016). The creep rate decreases to a value of minimum creep rate $\dot{\varepsilon}_{\mathrm{min}}$ and immediately increases after that minimum point. An exemplary creep curve for such a material is shown in Fig. 2.1 b) and marked with Ⓑ.

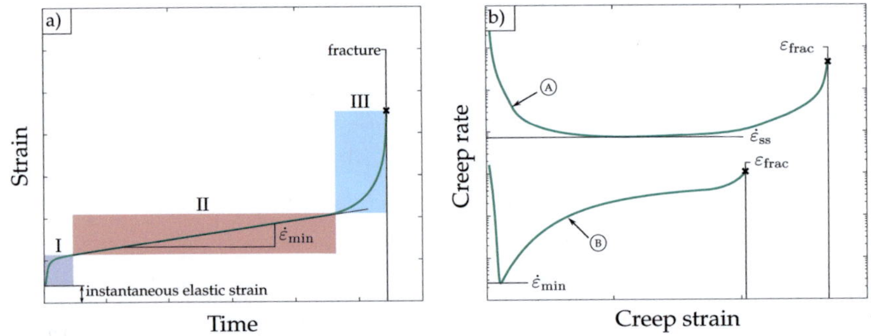

Figure 2.1: a) Exemplary creep curve with three typical creep regimes: the primary I, the secondary II, and the tertiary III creep regime. b) Creep rate versus creep strain for a material with a steady-state creep regime (creep curve marked with Ⓐ) and for a material without a steady-state creep regime (creep curve marked with Ⓑ).

Several creep mechanisms are discussed to explain the behavior of a material at elevated temperatures, e.g., the coarsening of dislocation subgrains (Blum, 2008) or of precipitates (Abe, 2009; Blum, 2008), the lattice diffusion (Nabarro, 1948; Herring, 1950), the grain boundary diffusion (Coble, 1963), the grain boundary sliding, e.g., (Nabarro and de Villiers, 1995), and the climb-plus-glide of dislocations, e.g., (Frost and Ashby, 1982; Nabarro and de Villiers, 1995). The controlling creep mechanism of the overall behavior depends on the material, the underlying microstructure, and the external creep conditions (e.g., the temperature and the applied stress).
For the characterization of the creep behavior of a material and the design of structures, the steady-state/minimum creep rate is an important characteristic value (Naumenko and Altenbach, 2016). Furthermore, the dependencies of the steady-state/minimum creep rate on stress and temperature are of primary interest and will be addressed in the following section. Additionally, the duration of each stage as well as the time and strain to fracture are important properties (Naumenko and Altenbach, 2016).

2.1.2 Temperature and stress dependent creep behavior

The minimum creep rate strongly depends on the applied stress and the temperature. Exemplary minimum creep rates for different applied stresses are shown in a full-logarithmic graph (Fig. 2.2 a)). For several materials a constant stress exponent $\bar{m}$ over a range of applied stresses can be observed. The value of the stress exponent is helpful to identify the creep mechanism, e.g., metallic materials possess a stress exponent in the range of 3 to 10 for climb-plus-glide based creep (Naumenko and Altenbach, 2016). The temperature dependency of the minimum creep rate is depicted exemplarily in a natural logarithmic versus T^{-1} graph (Fig. 2.2 b)). Analog to the stress exponent, a constant activation energy for creep $\bar{Q}_{\mathrm{C}}$ can be observed over a range of different temperatures.

For many materials, the dependency of the minimum creep rate to the applied stress as well as the temperature can, thus, be described by a product of two separate functions (Naumenko and Altenbach, 2016; Mukherjee, 2002). Several functions describing stress and temperature exist (Naumenko and Altenbach, 2016), whereas in the context of this work the power-law function for the stress and an Arrhenius function for the temperature were used, thus,

$$\dot{\varepsilon}_{\min} \propto \exp\left(-\frac{\bar{Q}_{\mathrm{C}}}{RT}\right)(\bar{\sigma})^{\bar{m}}. \tag{2.1}$$

Here, the temperature T is in Kelvin and R is the universal gas constant.

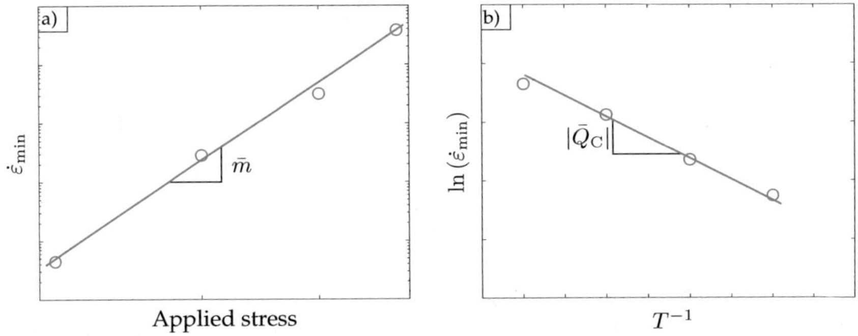

Figure 2.2: a) Minimum creep rate versus applied stress for a constant creep test temperature in a full-logarithmic representation. The slope of the fitted straight line corresponds to the stress exponent $\bar{m}$. b) Exemplary minimum creep rates showing the temperature dependency for a constant applied stress. The absolute value of the activation energy for creep $\bar{Q}_{\mathrm{C}}$ corresponds to the slope of the fitted straight line.

2.2 Directional solidification

2.2.1 General overview of the directional solidification

At high temperatures, grain boundaries can affect the creep resistance in an unfavorable way (e.g., by lattice diffusion (Nabarro, 1948; Herring, 1950) or grain boundary diffusion (Coble, 1963)). An overview of those weakening effects is given by Naumenko and Altenbach (2016). To avoid these grain boundaries, the directional solidification (DS) process can be employed to create a single-crystal material. During the DS process, a controlled solidification front moves through the sample which can be achieved by using, e.g., a Bridgman type (Misra et al., 1997) or an optical floating zone furnace (Bei and George, 2005). A schematical representation of an optical floating zone furnace is shown in Fig. 2.3 on the left. The upper and the lower rod are connected by the melt. To achieve a planar solidification front and a homogeneous melt, the upper and the lower rod are rotating in contrary directions. The solidification front moves from top to bottom by lowering both rods. For eutectic alloys, the directional solidification process can further be used to create a well aligned fibrous or lamellar microstructure (e.g., NiAl matrix reinforced by Cr fibers (Walter and Cline, 1970), Mo fibers (Bei and George, 2005), and Cr-Mo lamellas (Yu et al., 2014)). A review of DS intermetallic composites for high temperature structural applications is given by Bei et al. (2004). The fibers and the lamellas are aligned parallel to the growth direction (schematically depicted for a fibrous microstructure in Fig. 2.3 on the right), which leads to a strong enhancement of the creep resistance in growth direction.

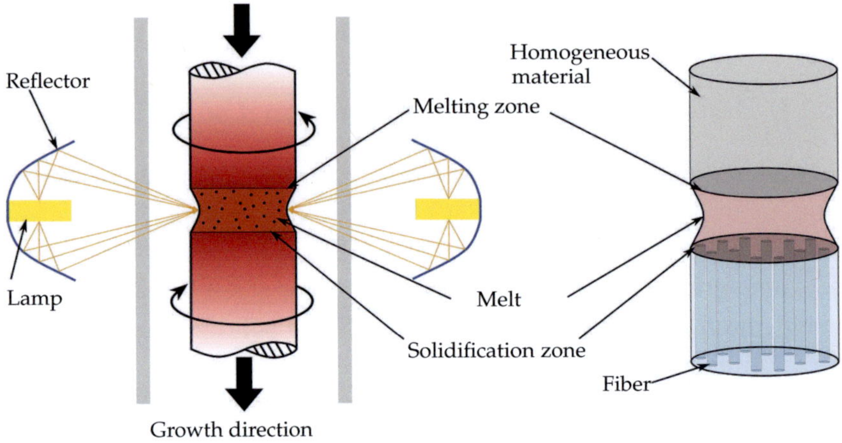

Figure 2.3: Schematical representation of an optical floating zone furnace on the left and the alignment of the fibers parallel to the growth direction due to the DS process on the right side.

2.2.2 Directionally solidified NiAl based eutectic alloys

Intermetallic compound NiAl

In the past, the intermetallic compound NiAl was a promising candidate for further applications at elevated temperatures (Haenschke et al., 2010; Frommeyer and Rablbauer, 2008; Darolia, 1994). This is due to (i) the high oxidation resistance, (ii) the high thermal conductivity as well as (iii) the higher melting temperature, and (iv) the lower density compared to currently used single-crystal nickel-based superalloys (Noebe et al., 1993; Miracle, 1993). The B2 crystal structure of NiAl can be described by one Al primitive cubic cell interpenetrating one Ni primitive cubic cell and is depicted in Fig. 2.4. This crystal structure exists over the composition range of 45-60 at.-% Ni (Noebe et al., 1993). Within this work, the compositions are given in atom-% (at.-%). The observed active slip systems of this crystal are given in Tab. 7.1 (Noebe et al., 1993; Miracle, 1993). Based on the slip systems,

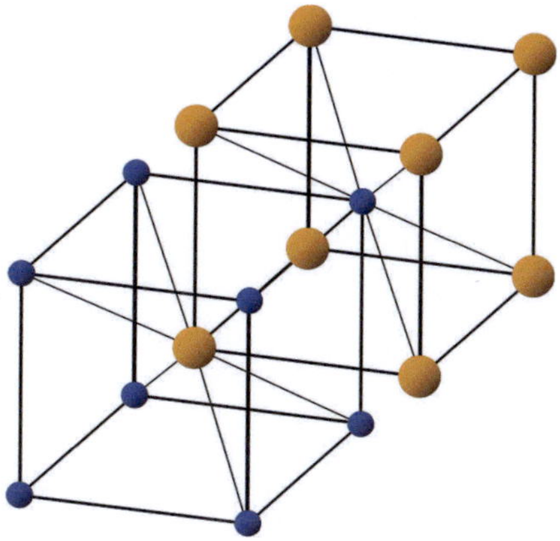

Figure 2.4: B2 crystal structure of NiAl consisting of one Al primitive cubic cell interpenetrating one Ni primitive cubic cell depicted by the blue and yellow atoms, respectively.

single-crystal NiAl exhibits two different types of slip behavior. The so called *soft*-orientation includes all non-$\langle 001 \rangle$ loading directions and has a dominant slip vector in the $\langle 001 \rangle$ direction. Contrary, the $\langle 001 \rangle$ *hard*-orientation exhibits a combination of $\langle 111 \rangle$, $\langle 110 \rangle$, and $\langle 100 \rangle$ slip vectors, depending on the temperature (e.g., Noebe et al. (1993)). Noebe et al. (1993) summarized the yield stress of near stoichiometric single-crystal NiAl for different orientations as a function of the temperature. The compound exhibits a relatively low ductile to brittle transition temperature (Haenschke et al., 2010), which is summarized for various compositions in Noebe et al. (1993). Frommeyer et al. (2010) determined the elastic constants for polycrystalline NiAl and Rusović and Warlimont (1977) used the ultrasonic pulse-echo method to reveal the elastic constants for a single-crystal.

However, the low creep resistance and the low room temperature fracture toughness currently limits the use of NiAl for high temperature structural applications, e.g., Noebe et al. (1993). The high temperature plastic flow behavior of stoichiometric NiAl was examined for a polycrystalline microstructure by Whittenberger (1987; 1988), for a single-crystal by Forbes et al. (1996) and for off-stoichiometric Ni-40Al by Whittenberger et al. (1991). Arzt and Grahle (1998) studied the creep behavior of oxide-dispersion strengthened NiAl. Nevertheless, all studied cases reveal the low creep resistance compared to currently used single-crystal nickel-based superalloys. Moreover, experimental measurements reveal that, in addition to the low creep resistance, the low room temperature fracture toughness of approx. $8\,\mathrm{MPa}\sqrt{\mathrm{m}}$ (Noebe et al., 1993; Miracle, 1993; Hu et al., 2012) constrains the use of NiAl for structural applications. Depending on the crystal orientation for single-crystals and on the grain size for polycrystals, the fracture toughness is summarized in Noebe et al. (1993); Miracle (1993). One possibility to simultaneously enhance the creep resistance and the room temperature fracture toughness is the employment of the directional solidification of eutectic alloys, where near-stoichiometric NiAl can be reinforced by a refractory metal, e.g., Cr, Mo, W, Re (Frommeyer and Rablbauer, 2008).

Directional solidified NiAl-Mo

Depending on the growth rate, the directional solidification of the eutectic NiAl-9Mo creates a well-aligned fibrous or cellular microstructure (Seemüller et al., 2013; Zhang et al., 2013). The minimum creep rate of well-aligned fibrous microstructures can be considerably lower than for cellular microstructures (Seemüller et al., 2013), making the former more interesting for high temperature structural applications. A more detailed analysis by Bei and George (2007) reveals that the chemical composition of the fibers is Mo-10Al-4Ni, whereas that of the matrix is Ni-45.2Al. In the context of this work, they will be referred to as fibers and matrix, respectively. The fibers are placed in a regular hexagonal arrangement (Zhang et al., 2013; Bei and George, 2005). Depending on the growth conditions, the area-equivalent fiber diameter of the DS NiAl-9Mo eutectic lies in the range of $80\,\mathrm{nm} - 900\,\mathrm{nm}$ (Seemüller et al.,

2013; Dudová et al., 2011; Bei and George, 2005; Zhang et al., 2013; Bei et al., 2005; Brady et al., 2014; Albiez et al., 2016a). Additionally, high aspect ratios of the Mo fiber length to fiber diameter were experimentally measured (> 200 in Bei et al. (2008a), > 100 in Haenschke et al. (2010), and ≈ 100 in Hu et al. (2013)). As a result of the DS process, a very good alignment of the fibers parallel to the growth direction was observed (Barabash et al., 2010a; Bei et al., 2008a) with the $\langle 100 \rangle$-orientation of the fibers as well as the matrix parallel to the growth direction (Dudová et al., 2011; Hu et al., 2012; Bei and George, 2005; Hu et al., 2013). A cube-on-cube orientation relation exists between these two phases with the interface boundary parallel to the $\{011\}$ crystal-plane (Hu et al., 2012; Bei and George, 2005; Hu et al., 2013; Joslin et al., 1995). Due to the eutectic composition, the volume fraction of the fibers is constrained to approx. $14\,\%$ (e.g., Bei et al. (2005); Bei and George (2005); Dudová et al. (2011)). A further advantage of the eutectic composition is the thermodynamical stability. As a result, the eutectic morphology stays stable under isothermal conditions (Johnson et al., 1995). Zhang et al. (2012) successfully produced a well-aligned fibrous microstructure with up to approx. $17.6\,\mathrm{vol.\text{-}\%}$ fibers, by the directional solidification of off-eutectic compositions. Following the Jackson and Hunt relation

$$\lambda^2 \vartheta = C_1, \tag{2.2}$$

the fiber spacing λ and the velocity of the solidification front ϑ are coupled by the constant C_1, which depends on the experimental conditions (Jackson and Hunt, 1966). Due to the volume fraction of the fiber c_{f} and the hexagonal arrangement, the fiber spacing is linked to the area-equivalent fiber diameter d by

$$\lambda = \sqrt{\frac{\pi d^2}{2\sqrt{3} c_{\mathrm{f}}}}. \tag{2.3}$$

A comparison of measured fiber diameters over growth rate is given in Albiez et al. (2016a).

The high temperature properties of DS NiAl-9Mo were investigated by bending and compression tests in Haenschke et al. (2010). Tension tests at elevated temperatures for various volume fractions were performed in Zhang et al. (2015) and for various fiber diameters in Hu et al. (2012); Bei and George (2005; 2007). Compression creep tests reveal a stress exponent of the composite $\bar{m} \approx 10\text{-}14$ (Albiez et al., 2016a; Dudová et al., 2011; Dudová et al., 2011) and an activation energy of $\bar{Q}_{\mathrm{C}} \approx 400\text{-}420\,\mathrm{kJ/mol}$ (Albiez et al., 2016a; Dudová et al., 2011), whereas for tension creep tests the stress exponent lies in the range of $\bar{m} \approx 3.5\text{-}5$ and the activation energy is $\bar{Q}_{\mathrm{C}} = 291 \pm 19\,\mathrm{kJ/mol}$ (Hu et al., 2013). However, all creep tests indicate the superior creep resistance of the DS NiAl-9Mo, where the minimum creep rate can be more than five orders of magnitude lower compared to the monolithic NiAl (Dudová et al., 2011).
The shape of the creep curve follows the exemplary creep curve of Fig. 2.1 (creep curve marked with Ⓑ), where after a sharp reduction of the creep rate in the primary creep regime the creep rate immediately increases after passing the minimum creep rate. Thus, the secondary creep regime with a distinct constant creep rate over creep strain is not visible. The room temperature fracture toughness is nearly doubled from $8\,\mathrm{MPa}\sqrt{\mathrm{m}}$ of [100]-oriented single-crystal NiAl to approx. $15\,\mathrm{MPa}\sqrt{\mathrm{m}}$ for the well-aligned [100]-oriented DS NiAl-9Mo (Misra et al., 1998; Misra and Gibala, 2000; Hu et al., 2012; Misra et al., 1997). Furthermore, Zhang et al. (2012) measured an even higher room temperature fracture toughness of $19.36\,\mathrm{MPa}\sqrt{\mathrm{m}}$ for a DS NiAl-16Mo material. No dominant toughening mechanism could be identified, therefore, the increased toughness is attributed to a combination of several mechanisms, such as crack trapping, crack bridging, and the intrinsic toughening of the NiAl matrix by generation of dislocations from the interface source (Misra et al., 1997; 1998). The ductile-to-brittle transition temperature was found to lie somewhere between $650\text{-}700\,^{\circ}\mathrm{C}$ (Bei and George, 2005; 2007). Brady et al. (2014) investigated the

oxidation behavior of the composite at dry and wet air. They concluded that the oxidation resistance increases with decreasing fiber diameter. The effect of rhenium (Re) addition was investigated by Misra et al. (1997; 1998), however, it leads to poor alignment of the fiber and, thus, to a loss of strength and toughness. Bei et al. (2005) investigated the thermal expansion of the fiber and the matrix parallel and transverse to the growth direction. Transverse to the growth direction, both phases expand independently over the entire range of temperature. However, parallel to the growth direction, the two phases co-expand up to $650\,^{\circ}\mathrm{C}$ with an average coefficient of thermal expansion of $12.8 \cdot 10^{-6}\,^{\circ}\mathrm{C}^{-1}$ (Bei et al., 2005). Beyond this temperature, the expansion of the fiber increases, which is a result of the free plastic flow of the NiAl matrix (Bei et al., 2005).

The yield strength of the fibers at room temperature (RT) was thoroughly investigated by Bei et al. (2007). They produced micro-pillars of the fiber by etching the matrix away and measured a yield strength close to the theoretical one (Bei et al., 2007). Such a high yield strength was also observed in the case of defect-free whiskers (Brenner, 1956). Hence, Bei et al. (2007) concluded that the fibers are essentially free of dislocations, which was later confirmed by others (Sudharshan Phani et al., 2011; Chisholm et al., 2012; Kwon et al., 2015). Further, by pre-straining the micro-pillar, Bei et al. (2008c) measured a decrease of the yield strength from $\approx 9.3\,\mathrm{GPa}$ at $0\,\%$ to $\approx 1\,\mathrm{GPa}$ at $11\,\%$ pre-strain for a constant fiber diameter of $500 - 550\,\mathrm{nm}$. This pronounced softening is depicted in Fig. 2.5 and indicates a transition from the defect-free behavior to a bulk material behavior (Bei et al., 2008c). A possible transition law will be discussed in Chapter 6. Similar results were obtained by tensile testing of the fiber material. Here, the effects of the considered volume of the fiber, thus, the amount of dislocations within tested fiber samples, were investigated and the results were explained by a Monte Carlo simulation (Johanns et al., 2011). Furthermore, the effects of the strain rate to the yield strength of the fiber were examined

(Loya et al., 2014). Dislocation measurements revealed a strong increase of the dislocation density with increasing pre-strain from $\approx 9 \cdot 10^8\,1/\mathrm{m}^2$ at $0\,\%$ to $\approx 2.5 \cdot 10^{14}\,1/\mathrm{m}^2$ at $16\,\%$ pre-strain (Sudharshan Phani et al., 2011). Additional measurements of the dislocation density in the fibers as a function of the pre-strain were done by Chisholm et al. (2012); Barabash et al. (2010a) and the results are shown together with the results of Sudharshan Phani et al. (2011) in Fig. 2.5. The description of the dislocation evolution will be discussed in Section 7.2.4. The inhomogeneous dislocation density distribution at medium pre-strain ($\approx$ 4-8 %) was explained by Sudharshan Phani et al. (2011) by cracks in the matrix as a result of the pre-straining or by strain localization due to strain softening of the fiber. Recent results, however, show the inhomogeneous activation of $\langle 111 \rangle$ slip in the fibers as a consequence of the stress due to the $\langle 100 \rangle$-type dislocations in the matrix, which are swept towards the interphase boundary with increasing pre-strain (Kwon et al., 2015). Barabash et al. (2010a) concluded that the deformation behavior of the composite material depends on the strain partitioning between both phases and that the fiber-matrix interface plays an important role. Furthermore, no elastic strain in the NiAl phase was measured above $650\,^\circ\mathrm{C}$, thus, the fiber phase essentially supports the entire applied load (Bei et al., 2008a;b). These results lead to the assumption that the strength of the fiber and the fiber-matrix interface controls the deformation behavior of the composite, whereas no interface debonding in the as-grown state was observed (Barabash et al., 2011).

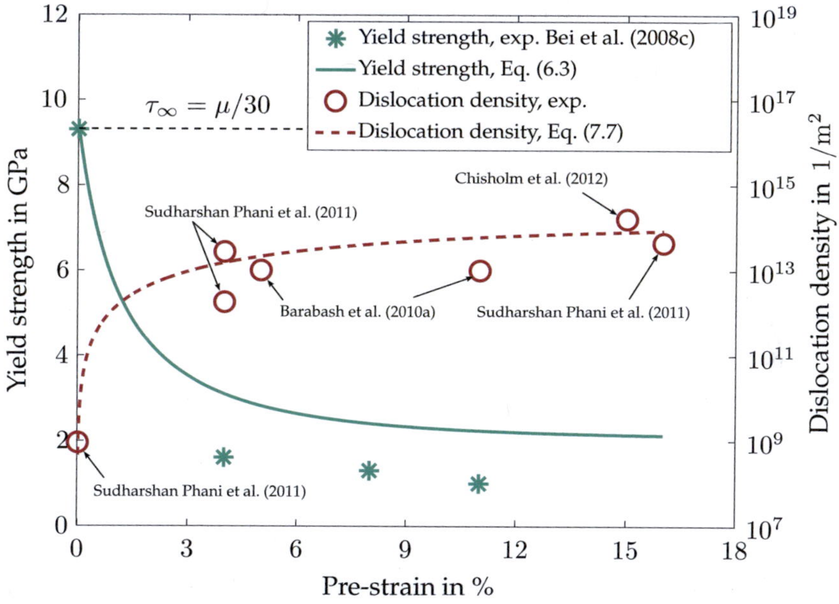

Figure 2.5: Yield strength (green curve) and dislocation density (red curve) as a function of the pre-strain. The dislocation densities were measured with TEM by Sudharshan Phani et al. (2011); Chisholm et al. (2012) and with a spatially resolved differential aperture X-ray microscopy by Barabash et al. (2010a). The measured values at 4 % pre-strain represent the maximum and minimum measured dislocation densities. Description of the dislocation evolution by the Kocks-Mecking dislocation evolution (Eq. (7.7)) is shown by the dashed line. The yield strength of the fibers was measured by Bei et al. (2008c) with micro-pillar experiments, revealing a yield strength close to the theoretical strength for the as-grown fibers. The continuous line shows the calculated strength by Eq. (6.3) and can be regarded as an upper limit. (Albiez et al. (2016a): Fig. 2)

Directional solidified NiAl-Cr(Mo)

The directional solidification of the eutectic NiAl-34Cr creates a well-aligned fibrous microstructure with ≈ 34 vol.-% Cr fibers (e.g., Misra and Gibala (2000)). Contrary to the DS NiAl-9Mo, the shape of the rods is non-faced (Misra and Gibala, 2000). Between the fibers and the matrix exists a cube-on-cube orientation, where the $\langle 001 \rangle$-orientation of both phases lies parallel to the growth direction (e.g., Misra and

Gibala (2000)). Walter and Cline (1970) examined the fiber spacing for several withdrawal rates and concluded that it follows the Jackson and Hunt relation in Eq. (2.2) (Jackson and Hunt, 1966). Misra and Gibala (2000) measured a yield strength of the composite at RT of $\approx 1200\,\mathrm{MPa}$ and observed for strains lower than $\approx 0.5\,\%$ only dislocations in the NiAl phase. However, at higher strains ($\approx 2\,\%$), also dislocations inside the fibers were observed (Misra and Gibala, 2000). Consequently, one may assume that the Cr fibers are essentially defect-free, similar to the previously discussed Mo fibers. The room temperature fracture toughness of $\approx 20\,\mathrm{MPa}\sqrt{\mathrm{m}}$ (Johnson et al., 1995) exceeds the one of pure NiAl and DS NiAl-9Mo. However, it may also depend on the directional process method (Misra and Gibala, 2000). The ductile to brittle transition temperature was found to be around $300\,^{\circ}\mathrm{C}$ (Haenschke et al., 2010). An improved creep resistance compared to monolithic NiAl was observed (e.g., Johnson et al. (1995); Haenschke et al. (2010)). In summary, the directional solidification of the NiAl-34Cr eutectic leads to a simultaneous improvement of the creep resistance and the room temperature fracture toughness compared to the monolithic NiAl.

Dependent on the withdrawal rate, the addition of Mo with more than $0.6\,\mathrm{at.\text{-}\%}$, however, results in a morphology change from the fibrous to a lamellar or cellular microstructure (Johnson et al., 1995). A lamellar/cellular microstructure is obtained for a low/high processing rate, respectively (Whittenberger et al., 1999). This morphology change can be favorable in terms of fracture resistance due to a superior fracture toughness of a lamellar microstructure compared to a fibrous microstructure (Chen et al., 2004). For the DS NiAl-31Cr-3Mo, Yu et al. (2014; 2016) measured a lamellar thickness of $\approx 800\,\mathrm{nm}$ and $\approx 400\,\mathrm{nm}$ for the NiAl layer and the Cr-based layer, respectively. Wang et al. (2018) investigated the Jackson and Hunt relationship for a DS NiAl-28Cr-6Mo compound. Similar to the previous case, a cube-on-cube orientation relationship was found (e.g., Chen et al. (2004)), however, for the lamellar microstructure, the $\langle 111 \rangle$-orientation of both phases is aligned parallel to the growth direction (e.g., Chen et al. (2004); Yu et al. (2016)). The lamellar microstructure before and after heat treatment was investigated by Wang et al. (2018) for different withdrawal rates and at different temperatures. They showed that the heat treatment has no significant effect on the microstructure produced by low withdrawal rates, whereas the stability becomes poor for high withdrawal rates (Wang et al., 2018). Yu et al. (2016) measured the thermal expansion behavior, where a co-expansion parallel and an independent expansion transverse to the growth direction was observed. Further, above $\approx 800\,^{\circ}\mathrm{C}$, the coefficient of thermal expansion of the NiAl matrix becomes larger than that of the lamella, indicating creep of the matrix at those temperatures (Yu et al., 2016). The room temperature fracture toughness was investigated for the DS NiAl-33Cr-1Mo by, e.g., Yang et al. (1997), for the DS NiAl-31Cr-3Mo by, e.g., Whittenberger et al. (1999); Johnson et al. (1995), and for the DS NiAl-28Cr-6Mo compound by, e.g., Johnson et al. (1995). Despite the different chemical compositions and different growth rates, all investigations reveal a superior fracture toughness compared to monolithic NiAl. Compression tests were conducted in the work of Whittenberger et al. (1999) at $1200\,\mathrm{K}$ ($\approx 930\,^{\circ}\mathrm{C}$) and $1300\,\mathrm{K}$ ($\approx 1030\,^{\circ}\mathrm{C}$) for specific strain rates. In addition, the yield strength was investigated by tensile tests at elevated temperatures in the work of Yang et al. (1997). The improved creep properties due to the directional solidification are shown for the DS NiAl-28Cr-6Mo compound in Johnson et al. (1995). The phase specific stress during compression was investigated by *in situ* neutron diffraction tests at room temperature in Yu et al. (2014) and up to $800\,^{\circ}\mathrm{C}$ in Yu et al. (2016). The stress in the Cr-based lamella was observed to be much higher than the ultimate strength of a single-phase bulk sample with the same chemical composition as for the lamella (Yu et al., 2014). Therefore, due to the similarities to the DS NiAl-9Mo, Yu et al. (2014) stated that the lamella may be essentially defect-free. However, the investigation in Yu et al. (2016) reveals strain hardening behavior in the matrix at relatively low temperatures and a strain softening at high temperatures, whereas the lamella hardens at all temperatures. To increase the volume fraction of the reinforcing phase, Shang et al. (2014) directionally solidified the hypereutectic alloy NiAl-32Cr-6Mo at different withdrawal rates. This investigation can be motivated by the fact that the strength of the composite is mainly controlled by the strength of the reinforcing phase (Johnson et al., 1995).

Chapter 3

Fundamentals of plastic deformation based on dislocations

3.1 Dislocation-induced plastic deformation

A metallic single crystal material consists of a highly ordered atomic structure. It is known from experimental work that such a material, loaded over a material-specific strength (yield strength), leaves the elastic state and starts to deform irreversibly. A schematical representation of this deformation is shown in Fig. 3.1.

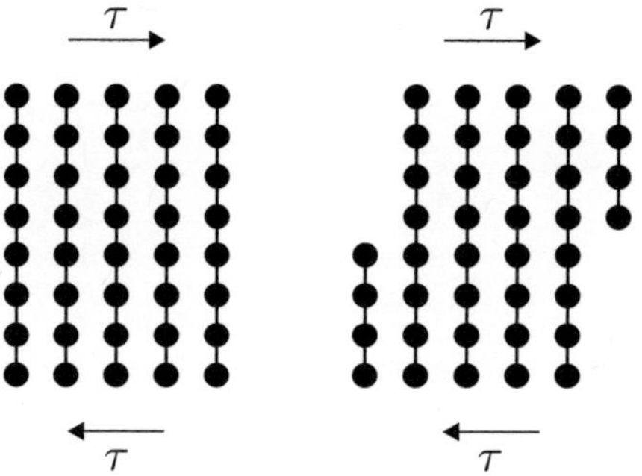

Figure 3.1: Theoretical plastic deformation without dislocations. The resolved shear stress τ shifts the total upper part of the crystal at once by one atom.

The necessary theoretical strength for this deformation was estimated to

$$\tau_\infty \approx \frac{\mu}{30} \tag{3.1}$$

by a crystal of bubbles (Cottrell, 1953). Here, μ stands for the material specific shear modulus. However, τ_∞ is immensely greater than the usually observed strength of metals ($\approx 10^{-4}\mu - 10^{-8}\mu$ (Hull and Bacon, 2011)). The striking difference between the theoretical strength and the experimentally observed strength was explained by the presence of dislocations (Orowan, 1934a;b;c; Taylor, 1934). A deformation based on dislocation glide allows a crystal to deform at significantly lower stresses and is schematically shown in Fig. 3.2. It is distinguishes between two basic types of dislocation geometry: an edge and a screw dislocation, depicted in Fig. 3.3 a) and c), respectively. An important characteristic of a dislocation is the Burgers vector $\boldsymbol{b}$, which can be determined by a Burgers circuit. A Burgers circuit is a closed atom-to-atom loop, depicted in Fig. 3.3 a) and c) for an edge and a screw dislocation, respectively. It is taken in a clockwise fashion, when looking along the dislocation line. By repeating the same atom-to-atom sequence in a perfect crystal (shown in Fig. 3.3 b) and d)), two options are possible.

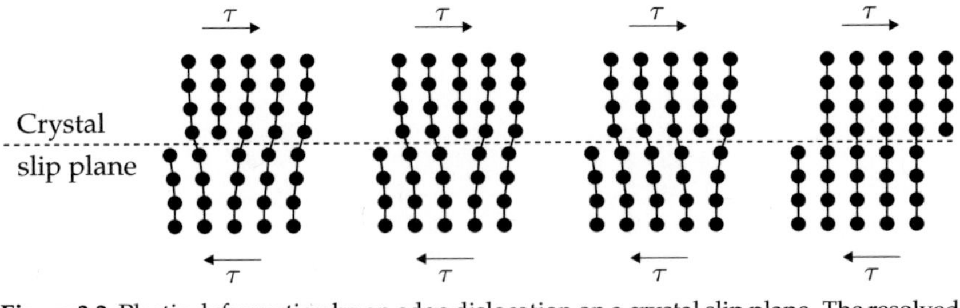

Figure 3.2: Plastic deformation by an edge dislocation on a crystal slip plane. The resolved shear stress τ for such a stepwise deformation is several times smaller compared to the theoretical strength.

In the first case, the loop in the perfect crystal is closed, meaning no Burgers vector exists. Whereas, in the second case, the loop in the perfect crystal does not close. The required vector to close the loop represents the Burgers vector $\boldsymbol{b}$. With the help of the Burgers vector, the following two rules can be used to distinguish between the edge and screw dislocation:

- The Burgers vector lies parallel to the line of a screw dislocation.
- The Burgers vector lies normal to the line of an edge dislocation.

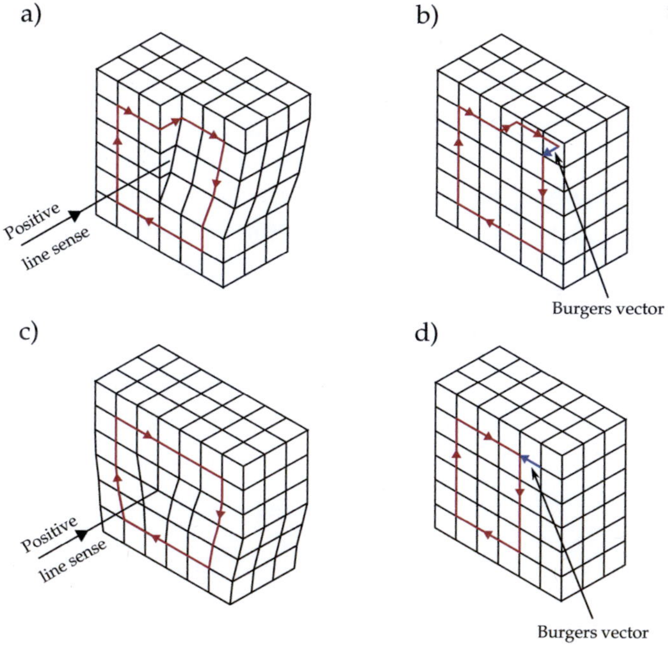

Figure 3.3: Burgers circuit around a) an edge dislocation and c) a screw dislocation as well as the same Burgers circuit in a perfect crystal depicted in b) and d), respectively.

In the most general case, however, the Burgers vector lies arbitrarily to the dislocation line. Further, it has a fixed length and direction for a single dislocation. At a meeting point of N_{dis} dislocations, the Burgers vector has to be conserved, thus,

$$\sum_{\alpha=1}^{N_{\mathrm{dis}}} \boldsymbol{b}_\alpha = \boldsymbol{0}. \tag{3.2}$$

Possible start and end points of dislocation lines are the surface of the crystal or grain boundaries, however, they can never start or end inside a crystal. The movement of dislocations can be classified into two basic types, glide and climb. Glide describes the movement in the surface which contains its dislocation line and the Burgers vector, whereas climb describes the motion out of this surface. The accumulation of several gliding dislocations results in slip, which is the most commonly observed plastic deformation (Hull and Bacon, 2011). It does not occur randomly within a crystal, but rather on specific slip systems, described by the normal vector of the slip plane $\boldsymbol{n}_\alpha$ and the slip direction $\boldsymbol{d}_\alpha$ of the slip system α. Thus, the slip system α can be described by a Schmid tensor $\boldsymbol{M}_\alpha = \boldsymbol{d}_\alpha \otimes \boldsymbol{n}_\alpha$. The average glide velocity v_α of dislocations in the slip system α depends on the applied stress via

$$v_\alpha \propto |\tau_\alpha|^m, \tag{3.3}$$

where τ_α and m correspond to the Schmid stress of the slip plane and the stress exponent, respectively, e.g., Hull and Bacon (2011). The Schmid stress can be calculated by

$$\tau_\alpha = \boldsymbol{\sigma} \cdot \boldsymbol{M}_\alpha \tag{3.4}$$

and corresponds to the projection of the Cauchy stress tensor $\boldsymbol{\sigma}$ to the slip system. By using Eq. (3.3) with the Orowan expression (Hirth and Lothe, 1992) for plastic slip

$$\dot{\gamma}_\alpha = \rho_\alpha b v_\alpha, \tag{3.5}$$

a relation between the plastic slip rate $\dot{\gamma}_\alpha$ and the Schmid stress of the slip system is achieved

$$\dot{\gamma}_\alpha \propto \rho_\alpha b |\tau_\alpha|^m. \tag{3.6}$$

Here, b is the absolute value of the Burgers vector and ρ_α the dislocation density in the slip system α, which is defined as the total length of the dislocation line per unit volume.

Two competing forces act on a dislocation which moves by glide. The glide force F_{glide}, acting on an element dl of the dislocation line of the Burgers vector $\boldsymbol{b}$ reads

$$F_{\text{glide}} = \tau b \mathrm{d}l \tag{3.7}$$

and results from the applied Schmid stress. It can be calculated by the work done by an element of a dislocation dl moving a unit distance (e.g., Hull and Bacon (2011)). In addition, a dislocation has a line tension which arises due to its strain energy. To minimize the total energy, the line tension intends to straighten the dislocation, where the induced force reads

$$F_{\text{tens}} = \bar{\alpha}\mu b^2, \tag{3.8}$$

with $\bar{\alpha} \approx 0.5$-1 (Hull and Bacon, 2011). To calculate the necessary Schmid stress τ to keep a dislocation curved with a radius of R_{dis}, an element of a dislocation shown in Fig. 3.4 is considered. For an angle $\mathrm{d}\varphi = \mathrm{d}l/R_{\text{dis}} \ll 1$, the force acting on the element due to the Schmid stress along the direction AB is $\tau_0 b \mathrm{d}l$.

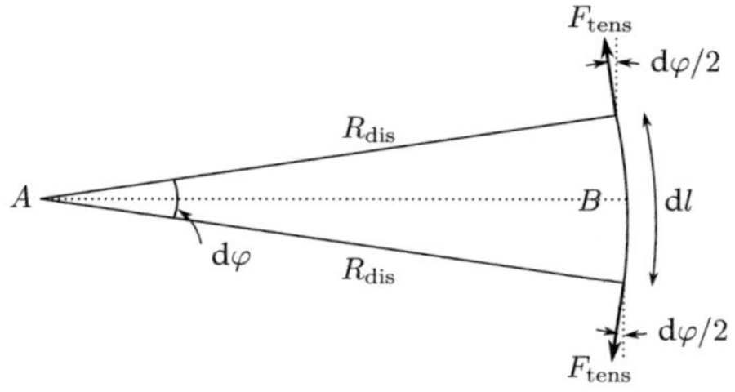

Figure 3.4: Curved dislocation-segment dl with radius R_{dis}. The line tension forces F_{tens} intend to straighten the dislocation in order to minimize the total energy (adapted from Hull and Bacon (2011)).

In the opposite direction, the force acting due to the line tension is $2F_{\mathrm{tens}} \sin(\mathrm{d}\varphi/2) \approx F_{\mathrm{tens}}\mathrm{d}\varphi$, for small $\mathrm{d}\varphi$. Thus, the two forces will be in equilibrium when

$$\tau \approx \frac{\bar{\alpha}\mu b}{R_{\mathrm{dis}}}. \tag{3.9}$$

This equation is crucial, since it can be used to explain the necessary stress to activate a dislocation source, which will be discussed in the next section. Further, it helps to explain several strengthening mechanisms as shown in Section 3.3.

3.2 Origin and multiplication of dislocations

The dislocation creation is seperated into two basic natures, the *homogeneous* nucleation, where the dislocation is created in a region of the crystal that is free of any defects, and the *heterogeneous* nucleation, where the dislocation is created with the help of defects. In the first case, the homogeneous nucleation, high stresses in the order of approximately $\mu/30$-$\mu/10$ are necessary (Hull and Bacon, 2011). Such high stresses were observed in the case of defect-free whiskers (Brenner, 1956) and as-grown Mo fibers (Bei et al., 2007), indicating a homogeneous creation of dislocations. However, generally dislocation creation appears by the movement and the multiplication of preexisting dislocations. This is based on the fact that the required stress is considerably lower and that most materials possess dislocations even in the as-cast state. Several heterogeneous dislocation multiplication mechanisms are possible and discussed in, e.g., Hirth and Lothe (1992); Hull and Bacon (2011). Therefore, only the relevant mechanisms for this work will be discussed in this section. A schematical representation of the *Frank-Read source* mechanism is depicted in Fig. 3.5 a) to d). The Burgers vector of the dislocation-segment AB lies in a slip plane and the dislocation is fixed at both points A and B (Fig. 3.5 a)). Possible reasons for such a fixation are for example dislocation intersections or precipitates. An applied Schmid stress forces the dislocation to bow out until the radius of curvature reaches its minimum point ($R_{\mathrm{min}} = L/2$), depicted in Fig. 3.5 b). As the dislocation expands further, it becomes unstable and grows until meeting itself (Fig. 3.5 c) to d)). Using the minimum radius of $R_{\mathrm{min}} = L/2$ and Eq. (3.9) leads to the necessary stress τ for this Frank-Read source mechanisms

$$\tau \approx \frac{\bar{\alpha}\mu b}{L}, \tag{3.10}$$

with L as spacing between the two fixing-points A and B (Hull and Bacon, 2011). Thus, this stress can be considerably lower than the theoretical strength due to $b/L \ll 1/30$.

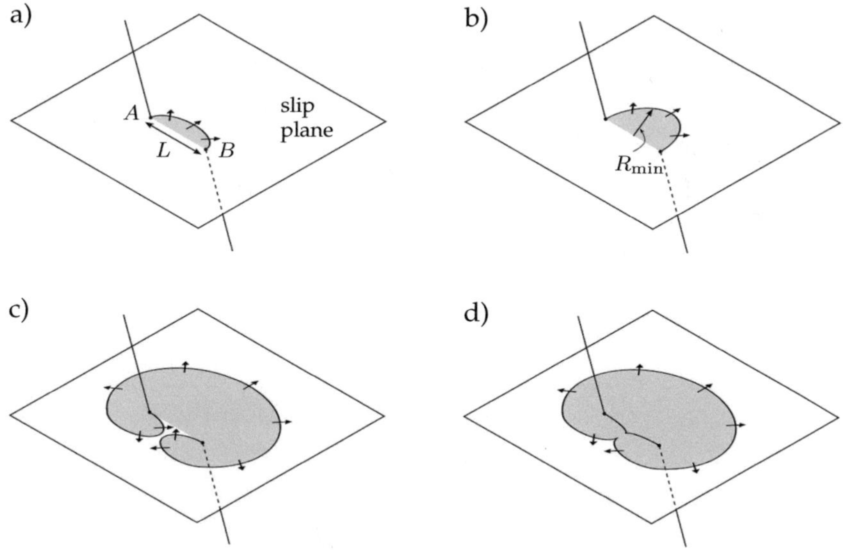

Figure 3.5: Schematical representation of the Frank-Read source mechanism, a) to d), following Hull and Bacon (2011). An applied Schmid stress bows the dislocation-segment which is fixed at both points A and B. Until the minimum radius ($R_{\mathrm{min}} = L/2$) is reached, the dislocation expansion is stable. The further expansion of the dislocation is unstable and it increases until meeting itself. The necessary stress for this mechanism is given by Eq. (3.10).

Another possible source of dislocations are grain boundaries in polycrystalline materials. This source can be activated by the large stress concentrations at the grain boundary due to pile-up stresses.

3.3 Dislocation strengthening mechanisms

The properties of a material depend, in general, on two competing different characteristic sizes. Arzt (1998) called these two length scales the *characteristic length* and the *size parameter*. The characteristic length describes the length of the physical phenomenon, thus, in the case of dislocations the absolute value of the Burgers vector, whereas the size parameter corresponds to a microstructural dimension. As discussed in Section 3.1, the equilibrium Schmid stress for a curved dislocation depends inversely on the radius of the dislocation (Eq. (3.9)). Therefore, one possibility to strengthen a material is to reduce the radius of the dislocation, which can be achieved by inserting obstacles into the material. A well-known strengthening mechanism, called *Orowan mechanism*, is the interaction between a dislocation and an array of hard particles, as depicted in Fig. 3.6 a). The particles are assumed impenetrable for dislocations and the bypassing stress (also called *Orowan stress*) can be calculated with the help of Eq. (3.9) to

$$\tau_{\mathrm{Or}} \approx \frac{\bar{\alpha}\mu b}{L}, \tag{3.11}$$

where L corresponds to the spacing between two particles. Therefore, the Orowan stress is equal to the Frank-Read source stress (Eq. (3.10)). Assuming a constant volume fraction of particles, the spacing is proportional to the radius of the particles R_{part}, thus, $L \propto R_{\mathrm{part}}$. Consequently, one would expect a monotonic increase of strength with a decreasing particle radius. However, experimental measurements indicate that a reduction of the particle size smaller than a critical radius $\hat{R}_{\mathrm{part}}$ leads to a loss in strength. This can be rationalized by considering the particles not as impenetrable, but rather with a certain strength against cutting. The cutting stress follows the relation $\tau_{\mathrm{cut}} \propto \sqrt{R_{\mathrm{part}}}$, therefore, it decreases with a decreasing particle radius (Arzt, 1998). The two dependencies are depicted in Fig. 3.6 b) and it follows that the maximal strength occurs for a particle radius $\hat{R}_{\mathrm{part}}$.

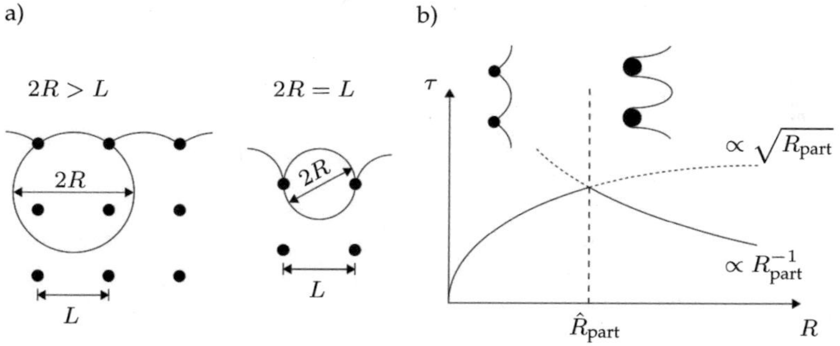

Figure 3.6: a) Schematical representation of the strengthening mechanism by an array of impenetrable particles (Orowan mechanism). The dislocation can overcome the particles for a maximal radius $2R = L$. b) In the case of penetrable particles, the cutting and the bypassing of the particles depend on its radius R_{part}, whereas the cutting stress scales with $\sqrt{R_{\mathrm{part}}}$ and the bypass stress with R_{part}^{-1}. Assuming a constant volume fraction of particles, the maximal strengthening occurs for a particle radius $\hat{R}_{\mathrm{part}}$.

An important and special case of the Orowan mechanism is the phenomenon of work hardening. In this case, the obstacles are forest dislocations, where the relevant size parameter corresponds to the average dislocation spacing, thus, $L = 1/\sqrt{\rho}$ with ρ as the dislocation density, e.g., Arzt (1998). With the help of Eq. (3.9), this leads to the classical *Taylor equation* for work hardening

$$\tau \approx \bar{\alpha}\mu b\sqrt{\rho}. \tag{3.12}$$

A further crucial size parameter which affects the plastic behavior in polycrystalline materials is the grain size. The classical *Hall-Petch* effect on the yield stress

$$\tau \propto \frac{1}{\sqrt{\bar{d}}} \tag{3.13}$$

can be explained by the pile-up of dislocations against grain boundaries (Hall, 1951; Petch, 1953). Here, $\bar{d}$ corresponds to the average grain diameter. However, as stated in, e.g., Arzt (1998); Zhu et al. (2008), several other possible explanations exist. Furthermore, in the case of thin films (film thickness $< 100\,\mathrm{nm}$ (Zhu et al., 2008)) and ultra-fine grained polycrystals, the Hall-Petch relation is no longer valid (reviewed in e.g., Kraft et al. (2010); Zhu et al. (2008)). This can be rationalized by the fact that at least one dislocation loop has to fit into a grain or into the film (Arzt, 1998). Therefore, the size parameter corresponds to the average grain diameter or the film thickness (also referred to as $\bar{d}$) and scales

$$\tau \propto \frac{1}{\bar{d}}. \tag{3.14}$$

This $\bar{d}^{-1}$ behavior was shown in the case of thin films by Arzt et al. (2001); Kraft et al. (2010); Zhu et al. (2008) and was motivated by the multiple activation of Frank-Read sources within the film by von Blanckenhagen et al. (2001; 2003). Nevertheless, for even smaller sizes, one observes a so-called inverse Hall-Petch effect, which can be described by

$$\tau \propto \frac{1}{\bar{d}} \ln\left(\frac{\bar{d}}{r_0}\right), \tag{3.15}$$

with r_0 as cut-off distance, e.g., Arzt (1998).

Chapter 4

Basics of continuum mechanics

4.1 Kinematics

Within continuum mechanics, the smooth movement of a material body $\mathcal{B}$ in time t through a three-dimensional euclidean space $\mathbb{R}^3$ is considered. The material body consists of material points and includes a singular interface $\mathcal{S}$ (depicted in Fig. 4.1). Further, it is limited by its boundary $\mathcal{A}$.

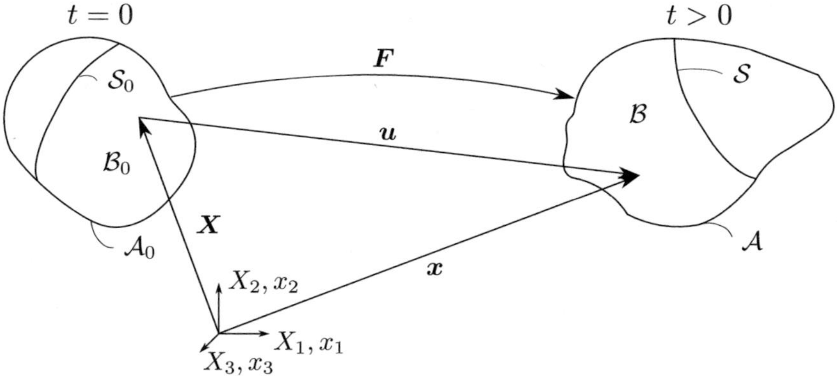

Figure 4.1: Representation of a material body ($\mathcal{B}_0$, $\mathcal{B}$) with its boundary ($\mathcal{A}_0$, $\mathcal{A}$) in the reference placement (for $t = 0$) and the current placement (for $t > 0$), respectively. The body contains a singular interface ($\mathcal{S}_0$, $\mathcal{S}$). The position of each material point can be described by $\boldsymbol{X}$ and $\boldsymbol{x}$ in the reference and the current placement, respectively.

The position of each material point is described by the vector $\boldsymbol{X}$ in the reference placement (for $t = 0$) and by the vector $\boldsymbol{x}$ in the current placement (for $t > 0$). The current position $\boldsymbol{x}$ can be specified as a function of $\boldsymbol{X}$ and t by

$$\boldsymbol{x} = \boldsymbol{\chi}(\boldsymbol{X}, t), \tag{4.1}$$

whereas the function $\boldsymbol{\chi}(\boldsymbol{X}, t)$ has to be continuous and differentiable, e.g., Bertram (2008). Respectively, the function

$$\boldsymbol{X} = \boldsymbol{\chi}^{-1}(\boldsymbol{x}, t) \tag{4.2}$$

describes the reference position $\boldsymbol{X}$ and depends on the current position as well as the time. The displacement vector $\boldsymbol{u}$ is defined as the difference of the position of a material point in the current and the reference placement, thus,

$$\boldsymbol{u} = \boldsymbol{x} - \boldsymbol{X} = \boldsymbol{\chi}(\boldsymbol{X}, t) - \boldsymbol{X}. \tag{4.3}$$

The velocity of a material point can be calculated by

$$\boldsymbol{v}_{\mathrm{L}}(\boldsymbol{X}, t) = \frac{\partial \boldsymbol{\chi}(\boldsymbol{X}, t)}{\partial t} \tag{4.4}$$

in the Lagrangian description. Replacing the argument $\boldsymbol{X}$ by $\boldsymbol{x}$ with help of Eq. (4.2) leads to the Eulerian description

$$\boldsymbol{v}_{\mathrm{E}}(\boldsymbol{x}, t) = \boldsymbol{v}_{\mathrm{L}}(\chi^{-1}(\boldsymbol{X}, t), t). \tag{4.5}$$

Both descriptions are, however, mathematically equivalent (Bertram, 2008) and the suffixes $(\cdot)_{\mathrm{L}}$ and $(\cdot)_{\mathrm{E}}$ will be dropped for brevity, if not explicitly necessary. In general, the time derivative of a field f of arbitrary order

$$\dot{f} = \left.\frac{\partial f_{\mathrm{L}}}{\partial t}\right|_{\boldsymbol{X}} = \left.\frac{\partial f_{\mathrm{E}}}{\partial t}\right|_{\boldsymbol{x}} + \frac{\partial f_{\mathrm{E}}}{\partial \boldsymbol{x}} \frac{\partial \boldsymbol{\chi}(\boldsymbol{x}, t)}{\partial t} = \frac{\partial f_{\mathrm{E}}}{\partial t} + \operatorname{grad}(f_{\mathrm{E}})\, \boldsymbol{v} \tag{4.6}$$

is called *material time derivative* and is indicated by a superimposed point. The spatial derivative used in the upper equation corresponds to the differential operator with respect to the Eulerian variable $\boldsymbol{x}$. Such differential operators will be indicated with a lower case first letter (e.g., $\mathrm{grad}\,(\cdot)$ or $\mathrm{div}\,(\cdot)$), whereas differential operators with respect to the Lagrangian variable $\boldsymbol{X}$ are denoted with a capitalized first letter (e.g., $\mathrm{Grad}\,(\cdot)$ or $\mathrm{Div}\,(\cdot)$). The deformation gradient

$$\boldsymbol{F} = \mathrm{Grad}\,(\boldsymbol{\chi}(\boldsymbol{X}, t)) \tag{4.7}$$

is important for the deformation analysis and can be used for the transformation of a material line element, an area element, and a volume element from the reference placement (upper case letter) to the current placement (lower case letter), shown in Box 4.1. A basic concept of deformation analysis is the decomposition of the deformation gradient into a rotation tensor ($\boldsymbol{R}$) and a stretch tensor (right $\boldsymbol{U}$ and left $\boldsymbol{V}$ stretch tensor), thus,

$$\boldsymbol{F} = \boldsymbol{R}\boldsymbol{U} = \boldsymbol{V}\boldsymbol{R}. \tag{4.8}$$

Box 4.1: Transformation of specific quantities between the reference and the current placement.

Line element	$\mathrm{d}\boldsymbol{x} = \boldsymbol{F}\mathrm{d}\boldsymbol{X}$
Area element	$\mathrm{d}\boldsymbol{a} = \det(\boldsymbol{F})\boldsymbol{F}^{-\mathsf{T}}\mathrm{d}\boldsymbol{A}$
Volume element	$\mathrm{d}v = \det(\boldsymbol{F})\mathrm{d}V = J\mathrm{d}V$

Several possibilities exist to describe the strain and are explained in, e.g., Bertram (2008). In this work, however, the Green strain tensor

$$\boldsymbol{E} = \frac{1}{2}(\boldsymbol{F}^{\mathsf{T}}\boldsymbol{F} - \boldsymbol{I}) \tag{4.9}$$

with $\boldsymbol{I}$ as identity tensor was used. The time rates of the deformation can be described by the spatial velocity gradient

$$\boldsymbol{L} = \mathrm{grad}\,(\boldsymbol{v}_{\mathrm{E}}) = \mathrm{Grad}\,(\boldsymbol{v}_{\mathrm{L}})\,\boldsymbol{F}^{-1} = \dot{\boldsymbol{F}}\boldsymbol{F}^{-1}, \tag{4.10}$$

which can be additively decomposed in a stretching tensor $\boldsymbol{D} = \mathrm{sym}(\boldsymbol{L})$ and a spin tensor $\boldsymbol{W} = \mathrm{skw}(\boldsymbol{L})$.

4.2 Balance equations

In this section, the balance equations of mass, linear momentum, angular momentum, total energy, and entropy will be introduced. Each balance equation is given with respect to the current configuration in its global and its local form. The balance equations are used to describe the motion of a material body $\mathcal{B}$ in time t and the evolution of the associated mechanical fields (e.g., mass density) (Hutter and Jöhnk, 2004). The mass balance of a material body with one component is given by

$$\frac{\mathrm{d}}{\mathrm{d}t}\int\limits_{\mathcal{V}} \varrho \mathrm{d}v = 0, \tag{4.11}$$

where ϱ stands for the mass density in the current placement. Here, it is assumed that no mass flux, no mass production, and no mass supply exist, e.g., Hutter and Jöhnk (2004). The local form of the mass balance and its jump condition are given by

$$\dot{\varrho} + \varrho \mathrm{div}\,(\boldsymbol{v}) = 0 \tag{4.12}$$

and

$$[\![\varrho(v_{\mathcal{S}}^{\perp} - v^{\perp})]\!] = 0, \tag{4.13}$$

respectively. Here, $v_{\mathcal{S}}^{\perp} = \boldsymbol{v}_{\mathcal{S}} \cdot \boldsymbol{n}_{\mathcal{S}}$ and $v^{\perp} = \boldsymbol{v} \cdot \boldsymbol{n}_{\mathcal{S}}$ are introduced with $\boldsymbol{v}_{\mathcal{S}}$ as the velocity of the singular interface Γ and $\boldsymbol{n}_{\mathcal{S}}$ as the normal vector of the singular interface.

Balancing the linear momentum of a material body yields

$$\frac{\mathrm{d}}{\mathrm{d}t}\int\limits_{\mathcal{V}} \varrho \boldsymbol{v} \mathrm{d}v = \int\limits_{\mathcal{A}} \boldsymbol{\sigma} \mathrm{d}\boldsymbol{a} + \int\limits_{\mathcal{V}} \varrho \boldsymbol{g} \mathrm{d}v \tag{4.14}$$

in the current placement with $\boldsymbol{\sigma}$ as the Cauchy stress. The first term on the right-hand side of Eq. (4.14) describes the flux of linear momentum, whereas the second term represents the supply of linear momentum with $\boldsymbol{g}$ as the force density per unit mass. No production of linear momentum exists, as it is conserved within a body. The local form and its jump condition are given by

$$\varrho\dot{\boldsymbol{v}} = \operatorname{div}(\boldsymbol{\sigma}) + \varrho\boldsymbol{g} \tag{4.15}$$

and

$$[\![\varrho\boldsymbol{v}(v^{\perp} - v_{\mathcal{S}}^{\perp}) - \boldsymbol{\sigma}\boldsymbol{n}_{\mathcal{S}}]\!] = 0, \tag{4.16}$$

respectively.
The balance of angular momentum of a non-polar continua in the current configuration is given by

$$\frac{\mathrm{d}}{\mathrm{d}t}\int\limits_{\mathcal{V}} \varrho\boldsymbol{x} \times \boldsymbol{v}\mathrm{d}v = \int\limits_{\mathcal{V}} \varrho\boldsymbol{x} \times \boldsymbol{g}\mathrm{d}v + \int\limits_{\mathcal{A}} \varrho\boldsymbol{x} \times \boldsymbol{\sigma}\mathrm{d}\boldsymbol{a} \tag{4.17}$$

and leads to $\boldsymbol{\sigma} = \boldsymbol{\sigma}^{\mathsf{T}}$ in its local form. Here, the symbol $\times$ represents the cross product of two vectors.
The balance of total energy

$$\frac{\mathrm{d}}{\mathrm{d}t}\int\limits_{\mathcal{V}} \varrho e + \frac{1}{2}\varrho\boldsymbol{v} \cdot \boldsymbol{v}\mathrm{d}v = \int\limits_{\mathcal{V}} \varrho\omega + \varrho\boldsymbol{g} \cdot \boldsymbol{v}\mathrm{d}v + \int\limits_{\mathcal{A}} (\boldsymbol{\sigma}\boldsymbol{v} - \boldsymbol{q}) \cdot \mathrm{d}\boldsymbol{a} \tag{4.18}$$

corresponds to the first law of thermodynamics (e.g., Hutter and Jöhnk (2004)). Here, the internal energy and the heat flux are denoted by e and $\boldsymbol{q}$, respectively. Further, $\varrho\omega$ corresponds to the energy supply, e.g., by heat sources and $\varrho\boldsymbol{g} \cdot \boldsymbol{v}$ to the supply of external volume forces. The local form of the balance of total energy and its jump condition are given by

$$\varrho(\dot{e} + \dot{\boldsymbol{v}} \cdot \boldsymbol{v}) = \varrho(\boldsymbol{g} \cdot \boldsymbol{v} + \omega) + \operatorname{div}(\boldsymbol{\sigma}\boldsymbol{v}) - \operatorname{div}(\boldsymbol{q}) \tag{4.19}$$

and

$$[\![\varrho(\frac{1}{2}\boldsymbol{v}\cdot\boldsymbol{v}+e)(v_{\mathcal{S}}^{\perp}-v^{\perp})+\boldsymbol{\sigma}\boldsymbol{v}\cdot\boldsymbol{n}_{\mathcal{S}}-\boldsymbol{q}\cdot\boldsymbol{n}_{\mathcal{S}}]\!]=0, \tag{4.20}$$

respectively. Simplifying the balance of total energy with the balance of linear momentum leads to the balance of internal energy

$$\varrho\dot{e}=\varrho\omega+\boldsymbol{\sigma}\cdot\boldsymbol{L}-\operatorname{div}(\boldsymbol{q})\,. \tag{4.21}$$

Here, $\boldsymbol{\sigma}\cdot\boldsymbol{L}$ can be interpreted as production, consequently, the internal energy e is not a conserved quantity.
The entropy is also not a conserved quantity, therefore, the balance of entropy includes entropy production terms and is given by

$$\frac{\mathrm{d}}{\mathrm{d}t}\int\limits_{\mathcal{V}}\varrho s\mathrm{d}v=\int\limits_{\mathcal{V}}\varrho\omega_{\mathrm{s}}+\varrho p_{\mathrm{s}}\mathrm{d}v-\int\limits_{\mathcal{A}}\boldsymbol{q}_{\mathrm{s}}\cdot\mathrm{d}\boldsymbol{a}+\int\limits_{\mathcal{S}}g_{\mathrm{s}}\mathrm{d}a \tag{4.22}$$

in the current placement, where s corresponds to the entropy. Here, $\varrho\omega_{\mathrm{s}}$ and $\boldsymbol{q}_{\mathrm{s}}$ stand for the entropy supply and its flux through the surface, respectively. In contrast to the previously introduced balance equations, the balance of entropy considers a surface entropy production g_{s} in addition to the production within the volume (p_{s}). For the mass balance and the balance of linear momentum, the missing surface production can be rationalized by considering the singular surface as massless. Furthermore, by stating the total energy to be a conserved quantity, including a surface production of total energy is not possible. The local form of the balance of entropy is given by

$$\varrho p_{\mathrm{s}}=\varrho\dot{s}+\operatorname{div}(\boldsymbol{q}_{\mathrm{s}})-\varrho\omega_{\mathrm{s}} \tag{4.23}$$

and its jump condition by

$$g_{\mathrm{s}}=[\![\boldsymbol{q}_{\mathrm{s}}\cdot\boldsymbol{n}_{\mathcal{S}}-\varrho s(v_{\mathcal{S}}^{\perp}-v^{\perp})]\!]. \tag{4.24}$$

The second law of thermodynamics states that for all permissible thermodynamic processes, entropy cannot be annihilated (e.g., Hutter and Jöhnk (2004)), thus,

$$p_\mathrm{s} \geq 0, \qquad g_\mathrm{s} \geq 0. \tag{4.25}$$

Substituting these inequalities (Eq. (4.25)) into the local form of the balance of entropy (Eq. (4.23)) and its jump condition (Eq. (4.24)) lead to the dissipation inequalities

$$\varrho p_\mathrm{s} = \varrho \dot{s} + \mathrm{div}\,(\boldsymbol{q}_\mathrm{s}) - \varrho \omega_\mathrm{s} \geq 0, \tag{4.26}$$

$$g_\mathrm{s} = [\![\boldsymbol{q}_\mathrm{s} \cdot \boldsymbol{n}_\mathcal{S} - \varrho s (v_\mathcal{S}^\perp - v^\perp)]\!] \geq 0 \tag{4.27}$$

in the current placement (e.g., Hutter and Jöhnk (2004)).

4.3 Principles of material theory

Within material theory, it is common to assume certain principles which help to define the form of a general constitutive equation more precisely. Following the *principle of determinism*, the stresses in a material point are only determined by the current and the past motion of the body, but not by its future one. Consequently, the principle of determinism reduces the temporal influence of deformations (Bertram, 2008). Different possibilities exist to take the past into consideration, whereas within this work, internal variables with evolution equations will be used. The spatial influence of deformations is restricted by the *principle of local action*, which claims that only a finite neighborhood of a material point influences its stress. In the limit case, thus, by considering the smallest possible neighborhood, the materials are called *simple materials*. Here, the stresses only depend on the motion of its infinitesimal neighborhood. The assumption of the *principle of material objectivity* (PMO) states that the Cauchy stress $\boldsymbol{\sigma}$ and the heat flux $\boldsymbol{q}$ are objective tensors (e.g., Bertram (2008); Hutter and Jöhnk (2004)). Consequently, if one observer identifies a constitutive equation for $\boldsymbol{\sigma}$ and $\boldsymbol{q}$, it can be used by any observer (Bertram, 2008). A further principle is the *principle of invariance under superimposed rigid body motions* (PISM) (Bertram, 2008). Following this principle, a rigid body motion only rotates the Cauchy stress. However, this principle is quite controversial due to counterexamples of, e.g., mechanics and thermodynamics (e.g., Bertram (2008); Krawietz (1986)). Nevertheless, for most materials under technical applications, it is accepted and applicable (Bertram, 2008). Finally, the *principle of form invariance* (PFI) states that the constitutive equations are invariant to a change of the observer (e.g., Bertram (2008)). It is noteworthy that two of the mentioned principles (PMO, PISM, or PFI) imply the third one.

4.4 Implications of the dissipation inequality

The equations introduced in Section 4.2 are quite general, however, they do not include any information about the materials (e.g., Hutter and Jöhnk (2004)). Furthermore, a determinate system can only be achieved by including constitutive equations. To ensure thermodynamical consistency, these constitutive equations should fulfill the second law of thermodynamics, thus, the dissipation inequalities given by Eq. (4.26) and Eq. (4.27). To identify indications of restrictions on the constitutive equations, the dissipation inequality in the reference placement is considered. Consequently, the balance of internal energy

$$\varrho_0 \dot{e} = \varrho_0 \omega + \boldsymbol{S} \cdot \dot{\boldsymbol{E}} - \operatorname{Div}\left(\boldsymbol{q}^{\star}\right) \tag{4.28}$$

with respect to the reference placement as well as the entropy balance

$$\varrho_0 p_{\mathrm{s}} = \varrho_0 \dot{s} + \operatorname{Div}\left(\boldsymbol{q}_{\mathrm{s}}^{\star}\right) - \varrho_0 \omega_{\mathrm{s}} \geq 0 \tag{4.29}$$

in the reference placement are necessary. They are derived from the corresponding balance equations (Eq. (4.21) and Eq. (4.26)) by using the transformations listed in Box 4.1. Here, $\boldsymbol{q}^{\star} = J\boldsymbol{F}^{-1}\boldsymbol{q}$ and $\boldsymbol{q}_{\mathrm{s}}^{\star} = J\boldsymbol{F}^{-1}\boldsymbol{q}_{\mathrm{s}}$

denote the fluxes $\boldsymbol{q}$ and $\boldsymbol{q}_\mathrm{s}$ with respect to the reference placement. In addition, ϱ_0 stands for the mass density in the reference placement with $\varrho_0 = J\varrho$. Furthermore, $\boldsymbol{S} = J\boldsymbol{F}^{-1}\boldsymbol{\sigma}\boldsymbol{F}^{-\mathsf{T}}$ corresponds to the second Piola-Kirchhoff stress and $\boldsymbol{E} = (\boldsymbol{F}^\mathsf{T}\boldsymbol{F} - \boldsymbol{I})/2$ to the Green strain tensor. The entropy flux is modeled by

$$\boldsymbol{q}_\mathrm{s}^\star = \frac{\boldsymbol{q}^\star}{T}, \tag{4.30}$$

following the classical thermodynamics (e.g., Hutter and Jöhnk (2004)). In Chapter 9, however, the entropy flux will be modified following the classical irreversible thermodynamics (e.g., Groot and Mazur (1984)). The entropy supply is assumed as

$$\varrho_0 \omega_\mathrm{s} = \varrho_0 \frac{\omega}{T}. \tag{4.31}$$

Introducing the free energy density

$$W = \varrho_0 \left(e - T\, s\right) \tag{4.32}$$

in the reference placement and using Eq. (4.28), Eq. (4.29), Eq. (4.30) as well as Eq. (4.31) leads to the inequality

$$\boldsymbol{S} \cdot \dot{\boldsymbol{E}} - \dot{W} - \varrho_0 s \dot{T} - \frac{\boldsymbol{q}^\star}{T} \cdot \mathrm{Grad}\,(T) \geq 0. \tag{4.33}$$

Considering the simplifications of the isothermal case with a homogeneous temperature distribution, the inequality reduces to

$$\boldsymbol{S} \cdot \dot{\boldsymbol{E}} - \dot{W} \geq 0. \tag{4.34}$$

The further constitutive equations used in this work are introduced in the following chapters, as they depend on the considered material and the specific physical problem. However, to ensure thermodynamical consistency, they all have to fulfill the inequality given by Eq. (4.29). Furthermore, the introduced principles (Section 4.3) are considered within the formulations of the constitutive equations.

4.5 General procedure for introducing micro-morphic variables

Forest (2009) proposed the *micromorphic approach* to introduce a general micromorphic variable $\check{\phi}$ as internal degree-of-freedom. This micromorphic variable can be a tensor variable of arbitrary rank and is associated with the state or internal variable ϕ (Forest, 2009). The approach leads to a general form of a partial differential equation used as equilibrium condition, which describes most strain gradient models in literature (Forest, 2009). The introduction of a penalty energy assures the coupling between the micromorphic variable and its associated variable.
The general form of the partial differential equation is derived in the finite deformation framework, however, it follows the approach of Forest (2009) in the infinitesimal deformation framework. Therefore, a body in its reference placement $\mathcal{B}_0$ and its current placement $\mathcal{B}$ containing a material singular interface (Γ_0, Γ) is considered (depicted in Fig. 4.2). Quantities in the reference placement are indicated by an additional $(\cdot)_0$ suffix. The consideration of a material singular interface simplifies the jump conditions discussed in Section 4.2 due to $\boldsymbol{v}_\mathcal{S} \cdot \boldsymbol{n}_\mathcal{S} = \boldsymbol{v} \cdot \boldsymbol{n}_\mathcal{S}$. The body is exposed to a macroscopic surface traction ($\bar{\boldsymbol{t}}_0$, $\bar{\boldsymbol{t}}$) and to a micromorphic surface traction ($\bar{\Xi}_{\mathrm{G}0}$, $\bar{\Xi}_{\mathrm{G}}$) at the external boundary $\mathcal{A}_{\mathrm{t}0}$, $\mathcal{A}_{\mathrm{t}}$ and $\mathcal{A}_{\Xi 0}$, $\mathcal{A}_{\Xi}$, respectively. Possible body forces acting on the body are neglected. The outward unit vector of the boundary is denoted by $\boldsymbol{n}$, whereas the orientation of the interface is described by the normal vector $\boldsymbol{n}_\Gamma$ pointing from $\mathcal{V}^-$ to $\mathcal{V}^+$. The displacement field $\boldsymbol{u}$ and its gradient are continuous through the interface, however, this does not necessarily count for the micromorphic variable $\check{\phi}$. Virtual quantities are denoted by $\delta(\cdot)$ and the virtual solution fields $\{\delta\boldsymbol{u}, \delta\check{\phi}\}$ are selected from the space of trial functions, thus, vanish at their respective Dirichlet boundaries.

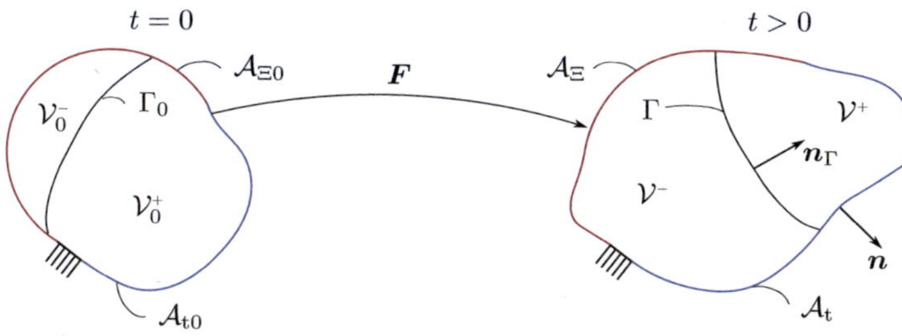

Figure 4.2: Body exposed to a macroscopic surface traction $(\bar{\boldsymbol{t}}_0, \bar{\boldsymbol{t}})$ at the external boundary $(\mathcal{A}_{t0}, \mathcal{A}_t)$ and to a micromorphic surface traction $(\bar{\Xi}_{G0}, \bar{\Xi}_G)$ at the external boundary $(\mathcal{A}_{\Xi 0}, \mathcal{A}_{\Xi})$ in the reference and the current placement, respectively. The unit vector $\boldsymbol{n}_\Gamma$ describes the orientation of the material interface.

Furthermore, the two virtual solution fields can be varied arbitrarily (Gurtin et al., 2010). The procedure can be sectioned as follows:

1. The internal variable ϕ which carries the targeted gradient effect is chosen. For sake of simplicity, it is treated as a scalar quantity. Further, the associated micromorphic variable $\check{\phi}$ is introduced, having the same tensor rank and same physical dimension as its corresponding internal variable (Forest, 2009). The micromorphic variable will be considered as an additional degree-of-freedom.
2. The virtual power of internal forces $\delta\mathcal{P}_{\text{int}}$ is extended by the power which is done by the micromorphic variable and its corresponding gradient, thus,

$$\begin{aligned}\delta\mathcal{P}_{\text{int}} = &\int\limits_{\mathcal{V}} \boldsymbol{\sigma}\cdot\delta\boldsymbol{L} + \pi_{\text{G}}\delta\dot{\check{\phi}} + \boldsymbol{\xi}_{\text{G}}\cdot\text{grad}\left(\delta\dot{\check{\phi}}\right)\,\text{d}v \\ &+ \int\limits_{\Gamma} \Xi_{\text{G}}^{+}\delta\dot{\check{\phi}}^{+} + \Xi_{\text{G}}^{-}\delta\dot{\check{\phi}}^{-}\text{d}a.\end{aligned} \tag{4.35}$$

   Here, the first integral term describes the internal virtual power in the bulk material with the symmetric Cauchy stress $\boldsymbol{\sigma}=\boldsymbol{\sigma}^{\mathsf{T}}$, the scalar micromorphic stress π_{G}, and the vector valued micromorphic stress $\boldsymbol{\xi}_{\text{G}}$. The second term considers the interface Γ with a micromorphic traction Ξ_G^+ conjugated to $\check{\dot{\phi}}^+$ and Ξ_G^- conjugated to $\check{\dot{\phi}}^-$. An alternative form of this integral reads

$$\int_\Gamma \Xi_G^+ \delta\check{\dot{\phi}}^+ + \Xi_G^- \delta\check{\dot{\phi}}^- \mathrm{da} = \int_\Gamma \Xi_G^s \langle\!\langle \delta\check{\dot{\phi}} \rangle\!\rangle + \Xi_G^j [\![\delta\check{\dot{\phi}}]\!] \mathrm{da} \tag{4.36}$$

with $\Xi_G^s = \Xi_G^+ + \Xi_G^-$ and $\Xi_G^j = 1/2(\Xi_G^+ - \Xi_G^-)$. This form offers a clear interpretation of the virtual power induced by the virtual jump $[\![\delta\check{\dot{\phi}}]\!]$ and the virtual mean value $\langle\!\langle\delta\check{\dot{\phi}}\rangle\!\rangle$ of the micromorphic variable.

3. Similar to the virtual power of internal forces, the virtual power of external forces $\delta\mathcal{P}_{\mathrm{ext}}$ is extended and has the form

$$\delta\mathcal{P}_{\mathrm{ext}} = \int_{\mathcal{A}_\mathrm{t}} \bar{\boldsymbol{t}} \cdot \delta\dot{\boldsymbol{u}} \mathrm{da} + \int_{\mathcal{A}_\Xi} \bar{\Xi}_G \delta\check{\dot{\phi}} \mathrm{da}. \tag{4.37}$$

4. The balance of the virtual power of internal and external forces, $\delta\mathcal{P}_{\mathrm{int}} = \delta\mathcal{P}_{\mathrm{ext}}$, leads to the necessary equilibrium condition in form of a partial differential equation, boundary conditions, and jump conditions, which are summarized in Box 4.2. The derivations are given in detail in Appendix A.1.

As stated and summarized by Forest (2009), the choice of the micromorphic variable and its associated internal variable depends on the studied physical problem (e.g., microdamage, microstrain gradient plasticity, or microdiffusion). Therefore, they are defined in the following sections depending on the considered investigations. Furthermore, the exact form of the generalized micromorphic stresses π_G, $\boldsymbol{\xi}_G$, Ξ_G^s, and Ξ_G^j will be derived and specified in the corresponding section by regarding the dissipation inequality.

Box 4.2: Equilibrium conditions, jump conditions, and boundary conditions for the displacement field $\boldsymbol{u}$ and a general micromorphic variable $\check{\phi}$.

Field variable	Equilibrium condition $\forall \boldsymbol{x} \in \mathcal{V}$	Jump condition on Γ	Boundary condition on $\mathcal{A}_\text{t} \vee \mathcal{A}_\Xi$
$\boldsymbol{u}$	$\text{div}\,(\boldsymbol{\sigma}) = \boldsymbol{0}$	$[\![\boldsymbol{\sigma}]\!]\boldsymbol{n}_\Gamma = \boldsymbol{0}$	$\boldsymbol{\sigma n} = \bar{\boldsymbol{t}}$
$\check{\phi}$	$\pi_\text{G} = \text{div}\,(\boldsymbol{\xi}_\text{G})$	$\Xi_\text{G}^\text{s} = [\![\boldsymbol{\xi}_\text{G}]\!] \cdot \boldsymbol{n}_\Gamma$, $\Xi_\text{G}^\text{j} = \langle\!\langle \boldsymbol{\xi}_\text{G} \rangle\!\rangle \cdot \boldsymbol{n}_\Gamma$	$\boldsymbol{\xi}_\text{G} \cdot \boldsymbol{n} = \bar{\Xi}_\text{G}$

4.6 Simplifications in case of infinitesimal deformations

The theory of infinitesimal deformations (also called *geometrically linear theory*) is a special case of the above introduced theory for finite deformations. The name geometrically linear theory results from the linearization of the strain tensors, e.g., Bertram (2008). It offers several advantages, especially in the case of readability and implementation, however, it is only valid in a small range of possible deformations. The norm of the displacement gradient $||\boldsymbol{H}|| = \sqrt{\boldsymbol{H} \cdot \boldsymbol{H}}$, with $\boldsymbol{H} = \boldsymbol{F} - \boldsymbol{I}$, is a possible measure for the deformation of a material line element. It contains the strain deformations and the rotations. For sufficiently small deformations, thus, $||\boldsymbol{H}|| \ll 1$, the theory of infinitesimal deformation conforms with the theory of finite deformation and, therefore, is applicable. As a result, no distinction between the position of a material point in the current placement ($\boldsymbol{x}$) and the reference placement ($\boldsymbol{X}$) is made. Consequently, the differential operators (e.g., $\text{grad}\,(\cdot)$ and $\text{Grad}\,(\cdot)$) are equal and are denoted by a lower case first letter. Finally, the material time derivative (Eq. (4.6)) reduces to $\dot{f} = \partial f / \partial t$.

Chapter 5

Numerical implementation

5.1 Time discretization

For the numerical integration of ordinary differential equations, e.g.,

$$\dot{y}(t) = g(y(t)), \quad y(0) = y_n \tag{5.1}$$

the so-called *generalized midpoint rule* can be used (Simo and Hughes, 1998). Here, $g : \mathbb{R} \to \mathbb{R}$ describes a smooth function and y represents the field variable. Depending on the parameter $\upsilon \in [0, 1]$, the generalized midpoint rule

$$y_{n+1} = y_n + \Delta t g(y_{n+\upsilon}), \tag{5.2}$$

$$y_{n+\upsilon} = \upsilon y_{n+1} + (1 - \upsilon) y_n \tag{5.3}$$

leads to the explicit Euler scheme for $\upsilon = 0$, the midpoint rule scheme for $\upsilon = 1/2$, and to the implicit Euler scheme for $\upsilon = 1$ (Simo and Hughes, 1998; Hairer et al., 1993). Here, y_{n+1} corresponds to the algorithmic approximation of the correct value $y(t_{n+1})$ at the time $t_{n+1} = t_n + \Delta t$ (Simo and Hughes, 1998). In this work, the numerical time discretization was conducted by the implicit Euler scheme, thus, $\upsilon = 1$. Furthermore, adaptive time step control was used during the simulations, whereas the maximal and minimal time steps were limited.

5.2 Spatial discretization

5.2.1 Motivation

In the context of the finite element method (FEM) a spatial discretization is necessary. Therefore, the body is divided into finite elements (also just called elements) which are connected by nodes, e.g., Fish and Belytschko (2007). The description of the field variable, for example the micromorphic variable $\check{\phi}$, within an element (e) is realized with shape functions $N_\beta^{(e)}$ of the nodes β and reads in the one-dimensional case

$$\check{\phi}(x) \approx \sum_{\beta=1}^{N_{\mathrm{el}}^{(e)}} N_\beta^{(e)}(x)\hat{\phi}_\beta. \tag{5.4}$$

Here, x represents the spatial coordinate, $\hat{\phi}_\beta$ is the nodal degree-of-freedom, and $N_{\mathrm{el}}^{(e)}$ corresponds to the number of nodes of the element (e). Several possible standard shape functions with different properties exist and the reader is referred to text books, e.g., Fish and Belytschko (2007) for further information. In the context of this work, however, non-standard finite elements were used due to the micromorphic approach and the possible discontinuity due to the interface as explained in Section 4.5. These non-standard finite elements consider the interface Γ and the additionally required degree-of-freedoms (DOF) to describe the micromorphic variables. In the case of a scalar micromorphic variable $\check{\phi}$, only one additional DOF is necessary. Even though, the displacement and its gradient are continuous through the interface, thus common shape functions would be sufficient, this is not necessarily the case for the additional micromorphic variable $\check{\phi}$, which might contain discontinuities at the interface.

A discontinuity can be classified into two basic types, namely *strong* and *weak* discontinuities. A field variable $\check{\phi}$ containing a strong discontinuity and a weak discontinuity is depicted in Fig. 5.1 for the one-dimensional case. An important research field containing strong and weak discontinuities at interfaces are crack simulations, thus, it was thoroughly explored in this context. The extended finite element method (XFEM) was introduced by Moës et al. (1999), where the discontinuous fields are incorporated by enriched shape functions. One of the advantages of this approach is the mesh-independent representation of the discontinuities (such as cracks), making this approach well-suited for moving interfaces (Belytschko et al., 2001). The generalized finite element method (GFEM) was presented by Strouboulis et al. (2000) as an alternative to the XFEM and was later extended to the interface-enriched generalized finite element method (IGFEM) by Soghrati et al. (2012). Nevertheless, as stated by Fries and Belytschko (2010), the XFEM and the GFEM method are merging and the differences between these two methods are vanishing.

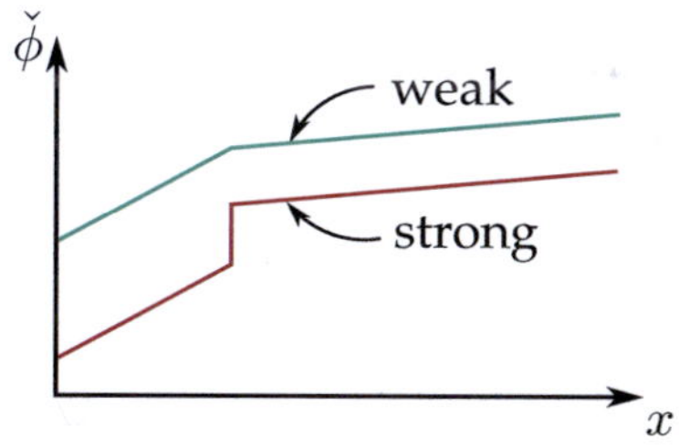

Figure 5.1: Schematical representation of a weak (green curve) and a strong (red curve) discontinuity of the field variable $\check{\phi}$.

Following the experimental results of Johnson et al. (1995), stating that the NiAl based eutectic morphology stays stable under isothermal conditions, the interfaces are regarded as fixed. Hence, no phase transformation and its resulting effects are accounted for, as observed, e.g., in the case of martensitic phase transformations (e.g., Bartel et al. (2011)). Additionally, only weak discontinuities of the micromorphic variable $\check{\phi}$ across the interfaces are considered, thus, the $\check{\phi}$-field is continuous, whereas its gradient field not necessarily. To profit from the advantages of a commercial software, the implementation of additional DOFs was realized within the ABAQUS UEL subroutine. The necessary shape functions and the numerical integration will be introduced in the following sections.

5.2.2 Finite element shape functions without discontinuities

To discretize the field variable within an element containing no interface, standard shape functions can be used. In the context of the micromorphic approach, however, the considered DOFs may be of different physical sort. For example, the micromorphic variable represents plastic slip in the gradient plasticity framework, e.g., Albiez et al. (2018); Wulfinghoff and Böhlke (2012), thus, the shape functions of the displacement should be one polynomial degree higher compared to the shape functions of the micromorphic variable. As a consequence, both, the plastic slip and the strain tensor $\boldsymbol{\varepsilon} = \mathrm{sym}(\mathrm{grad}\,(\boldsymbol{u}))$ (for infinite deformation) have the same polynomial degree. Therefore, it can be favorable to use different polynomial degrees for the DOFs. A quasi-isoparametric 20-node hexahedral element in the reference coordinate space (ξ, η, ζ) is depicted in Fig. 5.2. Quadratic shape functions consider all nodes, whereas the linear shape functions only account for the vertex nodes. This is similar to the work of Waffenschmidt et al. (2014), where the displacement field and an additional non-local damage variable was discretized with quadratic and linear shape functions, respectively.

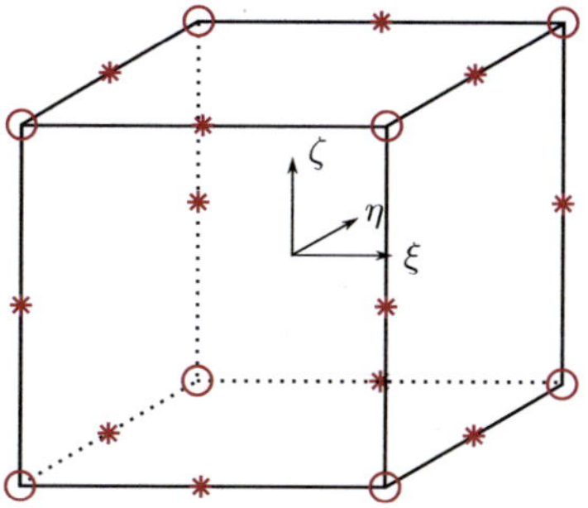

Figure 5.2: Quasi-isoparametric 20-node hexahedral element in the reference coordinate space (ξ, η, ζ). The position of the vertex nodes and the central nodes are indicated by the stars and by the circles, respectively. All nodes are considered for the quadratic shape functions, whereas only the vertex nodes are used for the linear ones.

5.2.3 Enrichment of shape functions for weak disontinuities

The discretization of field variables within an element containing an interface will be discussed in the following section. In this work, the displacement and its gradient are continuous through the interface, thus, common shape functions as described in the previous section (Section 5.2.2) can be used. However, a weak discontinuity is considered for the additional micromorphic variable $\check{\phi}$. Therefore, the enrichment of the shape functions in the context of the IGFEM will be introduced and explained for a one-dimensional problem. The idea of the IGFEM formulation was introduced by Soghrati et al. (2012) and is based on the introduction of nodes at the intersection of the interface and element edges, which possess generalized degrees-of-freedom.

The one-dimensional problem depicted in Fig. 5.3 is considered to explain the approach. Here, the field variable $\check{\phi}(x)$ should be discretized. The problem is discretized by three linear elements (e) and includes an interface Γ within the element (2) at $x = 0$. The field variable $\check{\phi}(x)$ can be discretized using standard shape functions $N_\beta^{(e)}(x)$ of the nodes β in

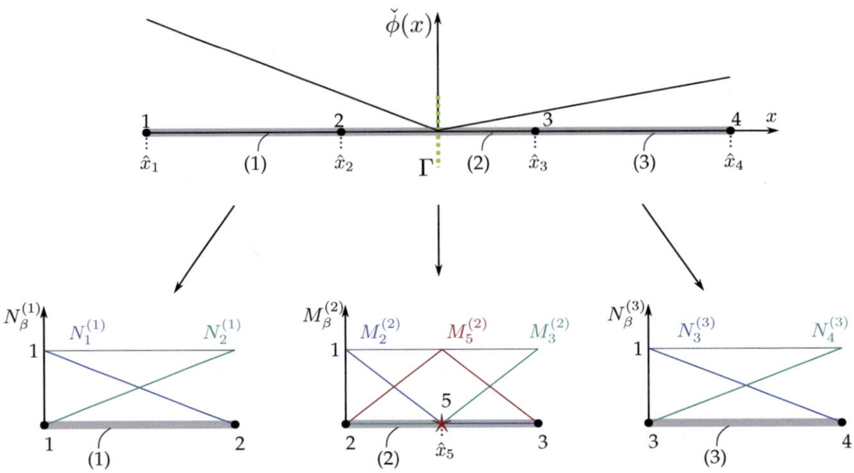

Figure 5.3: Discretization of the considered one-dimensional problem with a weak discontinuity of the field variable $\check{\phi}$ at the interface Γ. Following the IGFEM approach, the space is discretized by two elements ($e = 1, 3$) with standard shape functions $N_\beta^{(e)}(x)$ and one element ($e = 2$) with enriched shape functions $M_\beta^{(e)}(x)$. An additional node was created at the interface to consider the weak discontinuity (marked by the $\star$ symbol). (Albiez et al. (2018) Fig. 1)

elements without an interface. The weak discontinuity at the interface Γ is included by creating an additional node (marked by the $\star$ symbol in Fig. 5.3). Further, the shape functions of the element with the interface are enriched by this additional node. These enriched shape functions $M_\beta^{(e)}(x)$ read

$$\begin{aligned}
M_2^{(2)}(x) &= \frac{1}{2}\left(1 - \operatorname{sgn}\left(x - \hat{x}_5\right)\right)\left(\frac{x - \hat{x}_5}{\hat{x}_2 - \hat{x}_5}\right), \\
M_3^{(2)}(x) &= \frac{1}{2}\left(1 + \operatorname{sgn}\left(x - \hat{x}_5\right)\right)\left(\frac{x - \hat{x}_5}{\hat{x}_3 - \hat{x}_5}\right), \\
M_5^{(2)}(x) &= \frac{1}{2}\left(1 + \operatorname{sgn}\left(x - \hat{x}_5\right)\right)\left(\frac{\hat{x}_3 - x}{\hat{x}_3 - \hat{x}_5}\right) \\
&\quad - \frac{1}{2}\left(\operatorname{sgn}\left(x - \hat{x}_5\right) - 1\right)\left(\frac{\hat{x}_2 - x}{\hat{x}_2 - \hat{x}_5}\right),
\end{aligned} \tag{5.5}$$

where $\hat{x}_\beta$ represents the nodal coordinate. Further, they confirm the condition

$$\sum_{\beta=1}^{N_{\mathrm{el}}^{(e)}} M_\beta^{(e)}(x) = 1, \tag{5.6}$$

with $N_{\mathrm{el}}^{(e)}$ as the number of nodes of the element (e). As a result, the field variable $\check{\phi}(x)$ can be discretized over the considered space by

$$\check{\phi}(x) \approx \sum_{e \in I_\mathcal{V}} \left(\sum_{\beta=1}^{N_{\mathrm{el}}^{(e)}} N_\beta^{(e)}(x)\hat{\phi}_\beta \right) + \sum_{e \in I_\Gamma} \left(\sum_{\beta=1}^{N_{\mathrm{el}}^{(e)}} M_\beta^{(e)}(x)\hat{\phi}_\beta \right) . \tag{5.7}$$

Here, the sets of elements with and without an interface are represented by I_Γ and $I_\mathcal{V}$, respectively. Further, $\hat{\phi}_\beta$ corresponds to the nodal degree-of-freedom. Although the enriched shape functions $M_\beta^{(e)}(x)$ were introduced in the global coordinate system x, in this exemplary problem, they can also be defined in a reference coordinate system, thus, only one reference element is necessary. Exemplary enriched shape functions are shown in Fig. 5.4 in a two-dimensional plane, where the red marked nodes correspond to the additional interface nodes. The black nodes are the vertex nodes, whereas the grey nodes are necessary for quadratic shape functions (as discussed in Section 5.2.2). Consequently, the grey nodes are not essential for the enriched shape functions. The partition between the elements with and without an interface and, consequently, the enriched shape functions differ from the framework proposed by Soghrati et al. (2012). Nevertheless, the fundamental idea and the benefits of the IGFEM formulation remain. Furthermore, the introduced approach is favorable to include interface properties, which is necessary for the simulations described in Chapter 8.

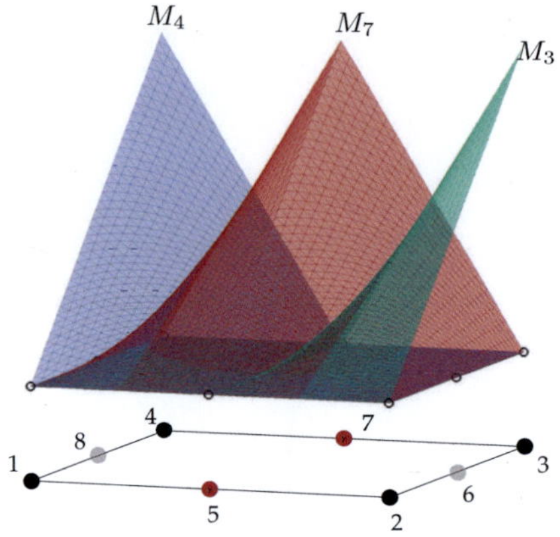

Figure 5.4: Exemplary enriched shape functions in a two-dimensional plane. The vertex nodes, the additional interface nodes, and the central nodes are marked by the black, the red, and the grey circles, respectively. The central nodes (grey circles) are necessary for the quadratic shape functions and are not considered for the enriched shape functions.

5.2.4 Numerical integration

The numerical integration is performed by a standard integration scheme

$$\int_{\mathcal{V}} g(x, y, z)\mathrm{d}v = J_{\mathrm{ref}} \sum_{\alpha=1}^{N_{\mathrm{IP}}} w_{\alpha}^{\mathrm{IP}} \cdot g(\xi^{\mathrm{IP}} - \alpha, \eta_{\alpha}^{\mathrm{IP}}, \zeta_{\alpha}^{\mathrm{IP}}), \tag{5.8}$$

where w_{α}^{IP} and $\xi_{\alpha}^{\mathrm{IP}}$, $\eta_{\alpha}^{\mathrm{IP}}$, $\zeta_{\alpha}^{\mathrm{IP}}$ correspond to the weight and the coordinates of the integration point α in the reference element with a total of N_{IP} integrations points. Further, J_{ref} corresponds to the determinant of the Jacobian matrix, which relates the physical coordinates (x, y, z) with the coordinates in the reference element (ξ, η, ζ) (Fish and Belytschko, 2007). For elements without an interface, the well-known weights and coordinates of the Gauss quadrature (also called Gauss-Legendre integration) were chosen. To reduce possible errors induced by the numerical integration, those elements include $N_{\mathrm{IP}} = 27$ integration points (IP) as depicted in Fig. 5.5 a). However, the standard Gauss quadrature in the weak form requires smoothness of the integrands (Fries and Belytschko, 2010), which is not necessarily given for the elements containing an interface. Thus, as a result of the discontinuity in the shape functions at the interface, the numerical integration demands special attention. Several possible approaches are explained and summarized by Fries and Belytschko (2010). In this work, the numerical integration for discontinuous shape functions is conducted by the decomposition of the elements, originially proposed by Moës et al. (1999). This approach is based on the decomposition of the elements along the discontinuity (thus along the interface) into subdomains, as depicted in Fig. 5.5 b). Within each subdomain, the smoothness of the integrand is ensured, thus, the standard Gauss quadrature can be applied for each subdomain. Therefore, each subdomain can be considered individually and 27 integration points are placed in each subdomain with the standard Gauss quadrature weights and positions. Furthermore, to consider the interface properties, additional integration points at the interface were introduced (Fig. 5.5 b)). Consequently, an element with an interface includes a total of 63 integration points (54 volume IP and 9 interface IP).

5.3 Periodic boundary conditions

The effective material response of heterogeneous materials can be estimated by numerical homogenization, besides others possibilities. Typically, not the complete geometry is discretized, but rather a representative part, called *representative volume element*. This representative part has to fulfill certain restrictions, such as the statistical homogeneity and ergodicity as well as the independence of boundary conditions (e.g., Hill (1963); Ostoja-Starzewski (2002)). However, the statistical homogeneity is obviously satisfied by considering a unit cell (UC) in the special case of a periodic microstructure. Exemplary cross-sections for unit cells with hexagonal and quadratic fiber arrangements are marked by the red rectangle and shown in Fig. 5.6 a) and b), respectively. In the case of an equal fiber spacing, the unit cell of the hexagonal arrangement is twice as large as the unit cell of the quadratic arrangement, making the quadratic arrangement numerically less expensive. Further, a unit cell of a lamella is depicted in Fig. 5.6 c). The independency of boundary conditions can be supported by the use of periodic boundary conditions, which are superior to homogeneous boundary conditions.

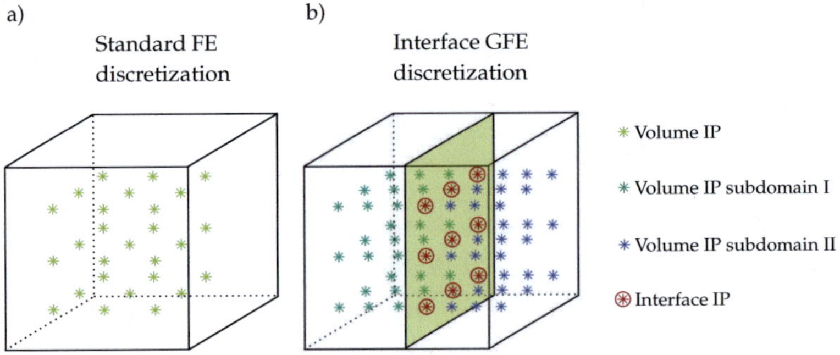

Figure 5.5: a) Standard finite element discretization for the numerical integration with 27 integration points. b) Decomposition of the element along the interface into two subdomains. The standard Gauss quadrature can be applied for each subdomain, whereas the interface properties are considered by the creation of additional integration points at the interface. (Albiez et al. (2018) Fig. 2)

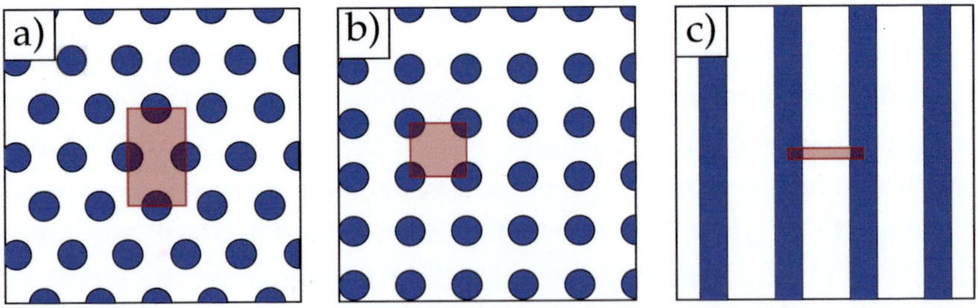

Figure 5.6: Exemplary unit cells of a) an hexagonal fiber arrangement, b) a quadratic fiber arrangement, and c) a lamellar microstructure. In each case, the unit cell is marked by the red rectangle.

In fact, the periodicity can get lost by applying homogeneous boundary conditions (e.g., Xia et al. (2003)). The basic idea of the periodic boundary condition is the coupling of the degree-of-freedom on the individual surfaces of the UC with each other. Here, the partition into *master* and *slaves* nodes is necessary and the coupling between those is forced by equations, summarized in Appendix A.2. They are explained in the case of the DOF for the displacement, however, they can be applied for any desired DOF.

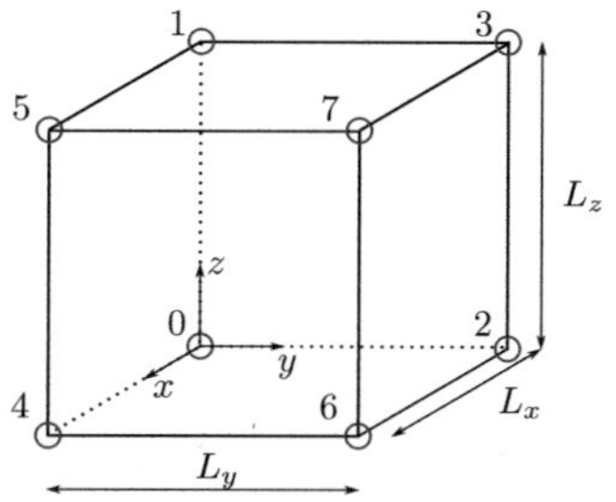

Figure 5.7: Node numbering for the independent nodal displacements as well as the dimensions L_x, L_y, and L_z of the rectangle.

Furthermore, the effective stress or strain field of the UC is realized by considering the DOF of specific *master*-nodes. To explain this approach, the rectangle with the dimensions L_x, L_y, and L_z in the x, y, and z direction is considered (depicted in Fig. 5.7). As a consequence of the symmetric property of the effective strain tensor $\boldsymbol{\varepsilon}^{\mathrm{eff}} = \mathrm{sym}(\mathrm{grad}\left(\boldsymbol{u}^{\mathrm{eff}}\right))$ with $\boldsymbol{u}^{\mathrm{eff}}$ as the effective displacement, the number of independent nodal displacements reduces to nine. It is, however, important to choose the nodal displacement such that any loading case can be obtained. Here, the displacements u_{x0}, u_{y0}, u_{z0}, u_{x1}, u_{y1}, u_{z1}, u_{x2}, $u^{y}{}_{2}$, and $u^{x}{}_{4}$ were chosen, whereas the first index corresponds to the axial direction of the displacement and the second index to the node number according to Fig. 5.7. Hence, the components of the effective strain tensor can be adjusted by

$$
\begin{aligned}
\varepsilon_{xx}^{\text{eff}} &= \frac{u_{x4} - u_{x0}}{L_x}, & \varepsilon_{xy}^{\text{eff}} = \varepsilon_{yx}^{\text{eff}} &= \frac{u_{x2} - u_{x0}}{L_y}, \\
\varepsilon_{yy}^{\text{eff}} &= \frac{u_{y2} - u_{y0}}{L_y}, & \varepsilon_{xz}^{\text{eff}} = \varepsilon_{zx}^{\text{eff}} &= \frac{u_{x1} - u_{x0}}{L_z}, \\
\varepsilon_{zz}^{\text{eff}} &= \frac{u_{z1} - u_{z0}}{L_z}, & \varepsilon_{yz}^{\text{eff}} = \varepsilon_{zy}^{\text{eff}} &= \frac{u_{y1} - u_{y0}}{L_z}.
\end{aligned}
\tag{5.9}
$$

Chapter 6

Transition law between different strengthening mechanisms

The strength of a single-phase material is usually described by the well-known square root Taylor hardening model (Eq. (3.12)). This hardening model is based on the self-interaction between individual dislocations and suggests direct proportionality between the strength and the dislocation density. Thus, one would expect a monotonic loss of strength with a decreasing dislocation density. However, experimental investigations reveal a significant increase in strength for a vanishing dislocation density, up to the theoretical strength τ_∞ in the case of an essentially dislocation free material (e.g., for defect-free whiskers (Brenner, 1956) and as-grown Mo fibers (Bei et al., 2007)). This increase of strength with decreasing dislocation density can be explained by the exhaustion of dislocations and dislocation sources, e.g., El-Awady (2015); Chisholm et al. (2012). Therefore, the Taylor hardening model is only valid in the case of a significantly high dislocation density, whereas the theoretical strength can only be achieved in the defect-free case. To bridge this gap between no dislocations and a significantly high dislocation density, a transition law is necessary (Lilleodden and Nix, 2006). A statistical model was proposed by Johnson and Ashby (1968), where the increase of strength with decreasing dislocation density was motivated by an increasing distance between jogs. They concluded that the strength of a material develops as

$$\begin{aligned} \tau^{\mathrm{F}} &\propto \rho^{-3/2} && \text{for} \quad \rho_{\mathrm{crit}} > \rho, \\ \tau^{\mathrm{F}} &\propto \sqrt{\rho} && \text{for} \quad \rho_{\mathrm{crit}} \leq \rho \end{aligned} \tag{6.1}$$

with the critical dislocation density ρ_{crit}. Thus, the strength increases with decreasing dislocation density as a result of the dislocation source strengthening mechanism for a dislocation density below a critical value. Furthermore, the strength increases with increasing dislocation density due to the Taylor hardening for densities above this critical value. A similar approach was conducted in Parthasarathy et al. (2007) by considering the stochastics of dislocation source lengths and by a Monte Carlo simulation considering the randomness of the Schmid factor by Sudharshan Phani et al. (2013). More recently, a generalized size-dependent, dislocation-based transition model which predicts the transition from dislocation-source strengthening to forest-dominated strengthening was proposed by El-Awady (2015). The dependency of the yield strength on the initial dislocation density ρ and the intrinsic size parameter d was found by evaluation of over 200 discrete dislocation dynamics simulations supported by micro-pillar experiments. The function which gives the best fit reads

$$\tau^{\mathrm{F}} = \frac{\tilde{\beta}\mu}{d\sqrt{\rho}} + \tilde{\alpha}\mu b\sqrt{\rho} \tag{6.2}$$

and is depicted in Fig. 6.1. Here, $\tilde{\alpha}$ as well as $\tilde{\beta}$ correspond to dimensionless constants and b is the length of the Burgers vector. However, it is well known that the strength of a defect-free material is limited by its theoretical strength $\tau_{\infty} \approx \mu/30$ (Cottrell, 1953), which can neither be fulfilled by the approach of Johnson and Ashby (1968) (Eq. (6.1)) nor by the approach of El-Awady (2015) (Eq. (6.2)). Therefore, the flow stress of a physical transition law should decrease starting from the theoretical strength for low dislocation densities and reach the normal square root Taylor hardening for high densities (as suggested by Lilleodden and Nix (2006)). A possible regularization of Eq. (6.2) was motivated in Albiez et al. (2016a) and reads

$$\tau^{\mathrm{F}} = \frac{\tau_\infty}{d\sqrt{\rho}+1} + \bar{\alpha}\mu b\sqrt{\rho}. \tag{6.3}$$

Here, $\tau_\infty = \mu/30$ stands for the theoretical strength, thus, all parameters have a physical meaning. In addition, the limit case $\lim_{\rho\to 0} \tau^{\mathrm{F}} = \tau_\infty$ with $\tau_\infty = \mu/30$ is fulfilled (Fig. 6.1). Nevertheless, the investigations by El-Awady (2015) have been performed in the range of $d\sqrt{\rho} \gg 1$ only, where both approaches (Eq. (6.2) and Eq. (6.3)) are similar (Fig. 6.1 for $\rho > 10^{14}\,1/\mathrm{m}^2$). Thus, the transition law proposed by Albiez et al. (2016a) is valid for an extended range and does not disagree with the results of El-Awady (2015).

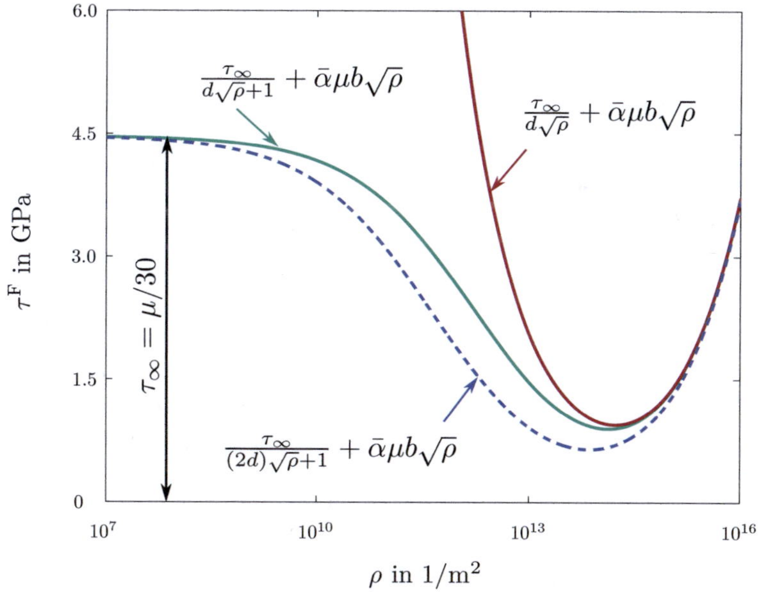

Figure 6.1: Comparison of the size-dependent transition models proposed by Albiez et al. (2016a) (green curve) and El-Awady (2015) (red curve). Contrary to the model of El-Awady (2015), the transition model of Albiez et al. (2016a) fulfills the limit of the theoretical strength for a defect-free crystal. In addition, the experimentally observed independence of the intrinsic size d for very low and very high dislocation densities is well reproduced (compare dotted blue curve with the continuous green one). (Albiez et al. (2016a) Fig. 1)

The transition law proposed by Albiez et al. (2016a) couples the size parameter d with the dislocation density. This coupling is crucial to explain several experimental findings. Usually, the yield strength dependency of the intrinsic size d is described by a model of type $\tau^{\mathrm{F}} \sim \tilde{k} d^{-n}$. Here, $\tilde{k}$ is assumed to be a material constant and one obtains the Hall-Petch type strengthening for $n = 0.5$. Nevertheless, $\tilde{k}$ collapses into a single value for different materials by using an inverse relationship, thus, $n = 1$ (Dunstan and Bushby, 2014). As discussed in Section 3.3, an inverse relationship was also found in the case of thin films by Arzt et al. (2001); von Blanckenhagen et al. (2001; 2003); Kraft et al. (2010); Zhu et al. (2008). However, as stated in Rinaldi (2011), the exponent n depends on the initial dislocation density. Such a dependency can be described by Eq. (6.3) and is depicted in Fig. 6.2. Here, τ^{F} is shown in a double-logarithmic plot over the size parameter d for two different initial dislocation densities. The negative slope of the curve corresponds to the exponent, which obviously depends on the combination of the size parameter and the initial dislocation density. For example, a lower initial dislocation density leads to a lower exponent at $d \approx 2 \cdot 10^{-5}\,\mathrm{m}^{-2}$ (Fig. 6.2). Thus, the coupling of the initial dislocation density with the exponent can be well described by Eq. (6.3). Furthermore, as shown for example by Bei et al. (2008c) in the micro-pillar experiments of Mo fibers, no size effect can be seen in case of the as-grown fibers (essentially dislocation free) as well as in case of the $11\,\%$ pre-strained fibers (dislocation density of $\approx 10^{14}\,\mathrm{m}^{-2}$ measured by Sudharshan Phani et al. (2011)). Consequently, the yield strength of a defect-free material is given by the theoretical strength and, thus, is independent of the size parameter d. Further, for materials containing a high enough dislocation density, the size parameter d is no longer dominating, but the average spacing $1/\sqrt{\rho}$ between individual dislocations is controlling the flow stress. These experimental findings are well described by the transition law proposed in Albiez et al. (2016a) as shown in Fig. 6.1 by the dashed line, which was calculated with a twice as high size parameter. Essentially no effects of the increased size parameter for low dislocation densities as well as high dislocation densities are visible.

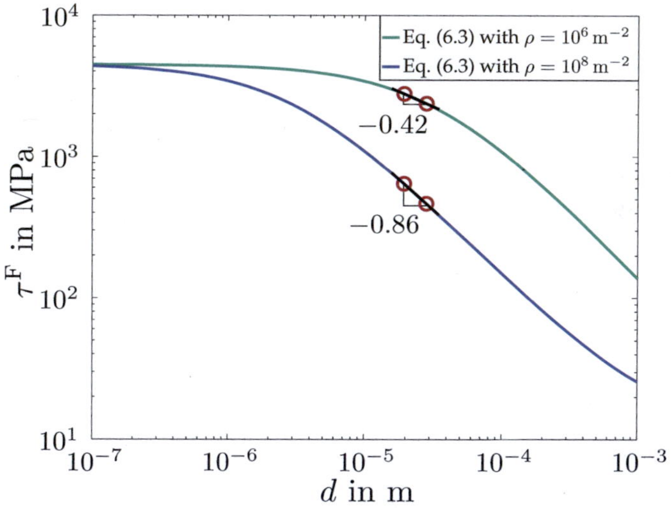

Figure 6.2: The transition model (Eq. (6.3)) describes the dependency of the exponent n on the initial dislocation density (based on experimental observations by Rinaldi (2011)).

In addition, the regime where an effect of the size parameter can be observed decreases for an increasing initial dislocation density. This is shown in Fig. 6.3, where the red circles indicate the limit of this regime. On the left side of the red circles, the size parameter has an effect, whereas on the right side of the circles, the effect of the size parameter vanishes. This is in good agreement with the work of Rinaldi (2011). Moreover, the experimentally observed softening of the Mo micro-pillars, as shown in Fig. 2.5, can be described by the transition law proposed in Albiez et al. (2016a). This calculated yield strength can be considered as an upper limit and describes the strong softening in the beginning with an asymptotical character for higher pre-strain, as observed by Bei et al. (2008c). Here, the dislocation density evolution shown in Fig. 2.5 together with the material properties of pure molybdenum listed in Tab. 7.1 were used. The applied dislocation density evolution will be discussed in Section 7.2.4. A similar trend of such a strong softening was observed by nanoindentation measurements, where the indentation load for the first pop-in event decreases with increasing pre-strain (reported for Mo in Bei et al. (2016), for Au wire in Guglielmi et al. (2018), and reviewed for several materials in Gao and Bei (2016)).

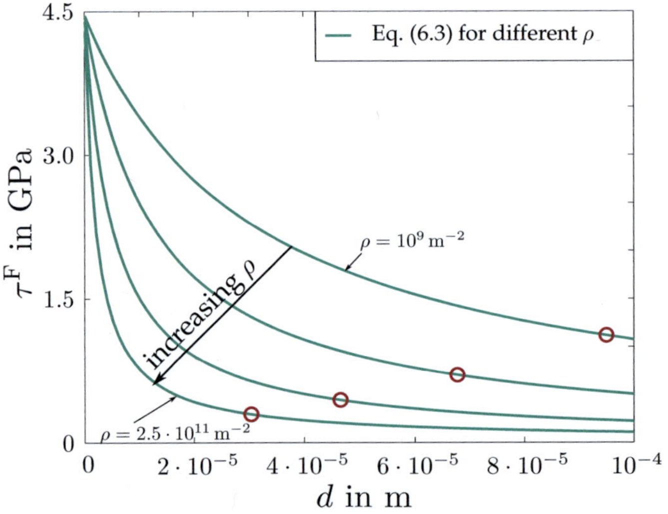

Figure 6.3: The size parameter d has an impact on the strength τ^F on the left side of the red circles, whereas the impact vanishes on the right side. The regime where an impact of the size parameter can be observed decreases for an increasing initial dislocation density.

In conclusion, the transition law proposed in Albiez et al. (2016a) fulfills the following experimental findings:

- The strength is limited by the theoretical strength in the case of a defect-free material.

- The effects of the size parameter for an essentially dislocation free material as well as for a material with a high dislocation density vanishes.
- Coupling between the dislocation density and the size parameter results in a nonconstant exponent n.
- Agreement with the results of El-Awady (2015), but also validity in an extended range.
- The regime where the size parameter affects the material's response depends on the dislocation density.
- Good agreement with the experimentally measured yield strength of Mo micro-pillars.

Chapter 7

Physically motivated crystal plasticity creep model for directionally solidified eutectics

7.1 Motivation

The reinforcement of near-stoichiometric NiAl with Mo fibers by directional solidification is of great interest for high temperature structural applications. This results from the increased creep resistance and the improved room temperature fracture toughness compared to the monolithic NiAl. Even though the fibers' behavior was explored intensively at room temperature and the creep behavior of the composite has been investigated at elevated temperatures several times (see Section 2.2.2), a physically motivated model which rationalizes the composite's behavior under various thermal and mechanical loads for different fiber diameters is still missing. There are some similarities between the short fiber reinforced Al matrix composites with the present NiAl-9Mo composite, however, the model introduced by Dlouhý et al. (1995) may not be valid here. This model describes a broad secondary creep range with fine oscillations as well as a tertiary creep regime due to fiber-breakage. Both phenomena, however, are not visible for the DS NiAl-9Mo.

In this chapter, a simple physical model proposed by Albiez et al. (2016a) will be presented. This model describes the creep behavior of the DS eutectic NiAl-9Mo and was validated by comparing simulated creep curves with experimentally measured ones (measured by I. Sprenger, C. Seemüller, and M. Heilmaier and published in Albiez et al. (2015; 2016a)). With the help of the simulation, the shape of the creep curve can be explained by considering the load transfer between the fiber and the matrix. Finally, the model and the insights due to the simulation will be discussed and summarized.

7.2 Balance equation and constitutive relationships

7.2.1 Balance equation

In the context of the simple physical model proposed in Albiez et al. (2016a), no micromorphic variable will be introduced. It follows that only the equilibrium condition of the displacement variable $\boldsymbol{u}$ has to be fulfilled (see Box 4.2). Thus, the considered conditions are given by

$$\begin{aligned} \mathbf{0} &= \operatorname{div}(\boldsymbol{\sigma}) \quad && \text{for} \quad \forall \boldsymbol{x} \in \mathcal{V}, \\ \bar{\boldsymbol{t}} &= \boldsymbol{\sigma}\boldsymbol{n} \quad && \text{on} \quad \mathcal{A}_{\mathrm{t}}. \end{aligned} \tag{7.1}$$

Further, no additional degree-of-freedom is necessary, allowing the implementation in the ABAQUS UMAT subroutine.

7.2.2 Basic constitutive relationships

Following Bertram (1998; 2005), the deformation gradient $\boldsymbol{F}$ is multiplicatively decomposed under the assumption of the existence of the elastic isomorphism ${\boldsymbol{F}_{\mathrm{p}}}^{-1} = \boldsymbol{P}$ into an elastic part $\boldsymbol{F}_{\mathrm{e}}$ and a plastic part $\boldsymbol{F}_{\mathrm{p}}$.

The decomposition is depicted in Fig. 7.1 and is given by

$$\boldsymbol{F} = \boldsymbol{F}_{\mathrm{e}}\boldsymbol{F}_{\mathrm{p}}. \tag{7.2}$$

The multiplicative combination of the mean rotation of the lattice $\boldsymbol{R}_{\mathrm{e}} \in Orth$ and the stretch of the lattice $\boldsymbol{U}_{\mathrm{e}} \in Sym$ leads to the elastic part of the deformation gradient, thus, $\boldsymbol{F}_{\mathrm{e}} = \boldsymbol{R}_{\mathrm{e}}\boldsymbol{U}_{\mathrm{e}}$. The plastic part $\boldsymbol{F}_{\mathrm{p}} \in Unim$ contains the slip caused by dislocation movement on slip systems. Under the assumption of small elastic strains, the St. Venant's elastic law

$$\boldsymbol{S}_{\mathrm{e}} = \mathbb{C}_0[\boldsymbol{E}_{\mathrm{e}}] \tag{7.3}$$

can be applied in the unloaded intermediate configuration. Here, the Green strain tensor is calculated by $\boldsymbol{E}_{\mathrm{e}} = (\boldsymbol{F}_{\mathrm{e}}{}^{\mathsf{T}}\boldsymbol{F}_{\mathrm{e}} - \boldsymbol{I})/2$ with $\boldsymbol{I}$ as identity tensor and $\mathbb{C}_0$ denotes the stiffness tensor. The plastic deformation is based on the movement of dislocations, consequently, the plastic deformation process is assumed as isochoric, i.e. $J = \det(\boldsymbol{F}) = \det(\boldsymbol{F}_{\mathrm{e}})$ and $\det(\boldsymbol{F}_{\mathrm{p}}) = 1$.

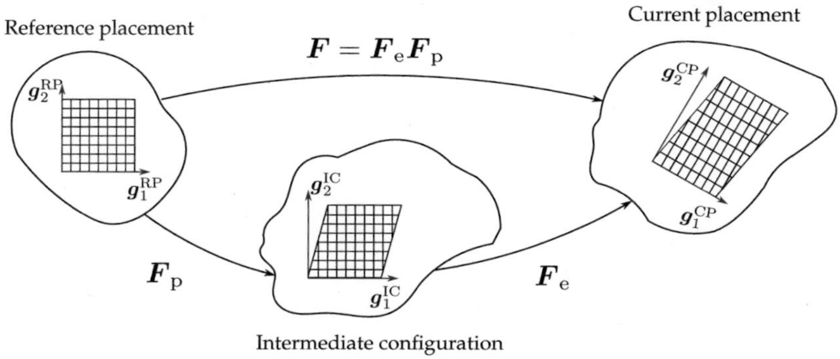

Figure 7.1: Schematical illustration of the multiplicative decomposition of the deformation under the assumption of the existence of the elastic isomorphism.

Further, the total velocity gradient $\boldsymbol{L} = \dot{\boldsymbol{F}}\boldsymbol{F}^{-1}$ and the Cauchy stress

$$\boldsymbol{\sigma} = J^{-1}\boldsymbol{F}_{\mathrm{e}}\boldsymbol{S}_{\mathrm{e}}\boldsymbol{F}_{\mathrm{e}}^{\mathsf{T}} \tag{7.4}$$

are defined in the current placement. In contrast, the plastic velocity gradient is obtained by $\boldsymbol{L}_{\mathrm{p}} = \dot{\boldsymbol{F}}_{\mathrm{p}}\boldsymbol{F}_{\mathrm{p}}^{-1}$ in the unloaded intermediate configuration and is associated to the slip rate $\dot{\gamma}_\alpha$ of mobile dislocations on the slip plane α such that

$$\boldsymbol{L}_{\mathrm{p}} = \sum_{\alpha=1}^{N_{\mathrm{GS}}} \dot{\gamma}_\alpha \tilde{\boldsymbol{M}}_\alpha. \tag{7.5}$$

Here, the Schmid tensor $\tilde{\boldsymbol{M}}_\alpha$ of the slip system α is defined in terms of the constant slip direction $\tilde{\boldsymbol{d}}_\alpha$ and the constant slip plane normal $\tilde{\boldsymbol{n}}_\alpha$ in the intermediate configuration and reads $\tilde{\boldsymbol{M}}_\alpha = \tilde{\boldsymbol{d}}_\alpha \otimes \tilde{\boldsymbol{n}}_\alpha$. The total number of slip systems N_{GS} depends on the material and is summarized in Tab. 7.1. As stated in Section 3.1, the plastic slip rate $\dot{\gamma}_\alpha$ of the slip system α is governed by the corresponding Schmid stress τ_α in a power-law way (compare Eq. (3.6)). Consequently, the slip rate is given by the temperature-compensated power-law equation

$$\dot{\gamma}_\alpha = \dot{\gamma}_0 \mathrm{sgn}\left(\tau_\alpha\right) \exp\left(-\frac{Q_{\mathrm{C}}}{RT}\right)\left(\frac{|\tau_\alpha|}{\tau_\alpha^{\mathrm{F}}}\right)^m \tag{7.6}$$

with the stress exponent m, the reference shear rate $\dot{\gamma}_0$, and the hardening variable τ_α^{F}. Further, the Arrhenius term in Eq. (7.6) describes the dependency of the slip rate on the temperature. Here, T corresponds to the creep test temperature in Kelvin, Q_{C} is the activation energy for creep, and R is the universal gas constant. For a creep mechanism limited by climb of edge dislocations, the activation energy for creep Q_{C} corresponds approximately to the activation energy of self-diffusion Q_{SD}, e.g., Raj (2002). An isotropic hardening approach $\tau_\alpha^{\mathrm{F}} = \tau^{\mathrm{F}}$ is applied due to the difficult experimental identification of the hardening behavior of each slip system (Bertram, 2005). The hardening variable τ^{F} depends, however, on the total dislocation density $\rho = \sum_{\alpha=1}^{N_{\mathrm{GS}}} \rho_\alpha$ and will be discussed in the following section.

7.2.3 Hardening model of the matrix and the fibers

Hardening model of the matrix

The elastic strain in the NiAl matrix was measured by tensile testing of the composite at $800\,°\mathrm{C}$ by Bei et al. (2005) with *in situ* neutron diffraction. They observed the absence of elastic strain in the matrix, revealing its free plastic flow at high temperatures. This was confirmed by further investigations in Bei et al. (2008a;b). In addition, no significant hardening of [100]-oriented single-crystal Ni-40Al was observed by Whittenberger et al. (1991) during compression testing at $1100\,\mathrm{K}$ ($\approx 830\,°\mathrm{C}$). Consequently, an ideal plastic behavior $\tau^{\mathrm{F}} = \tau_0$ was used to model the plastic deformation of the matrix.

Hardening model of the fibers

The fiber size of the DS eutectic NiAl-9Mo depends on the growth conditions and lies in the range of $80\,\mathrm{nm} - 900\,\mathrm{nm}$ (see Section 2.2.2). For this reason, a hardening evolution accounting for the size effects is required to model the hardening behavior of the fibers. Furthermore, Bei et al. (2008c) observed a dramatic loss in strength due to pre-straining. They measured a decrease of the yield strength from $\approx 9.3\,\mathrm{GPa}$ at $0\,\%$ to $\approx 1\,\mathrm{GPa}$ at $11\,\%$ pre-strain for a constant fiber diameter of $500 - 550\,\mathrm{nm}$. Further investigations by Sudharshan Phani et al. (2011) revealed an increase of the dislocation density with increasing pre-strain (from $\approx 9 \cdot 10^8\,1/\mathrm{m}^2$ at $0\,\%$ to $\approx 2.5 \cdot 10^{14}\,1/\mathrm{m}^2$ at $16\,\%$ pre-strain). Both investigations indicate a transition from theoretical strength to dislocation-based strengthening mechanisms. It follows that the hardening of the fiber material can be described by the transition law proposed by Albiez et al. (2016a), which was discussed in detail in Chapter 6.

7.2.4 Dislocation density evolution in the fibers

Results of experimental dislocation measurements of Sudharshan Phani et al. (2011); Chisholm et al. (2012); Barabash et al. (2010a) are summarized in Fig. 2.5. The as-grown fibers, essentially, do not possess any defects (Sudharshan Phani et al., 2011), which explains the measured yield strength near to the theoretical one (Bei et al., 2007). The dislocation density increases by pre-straining the fibers, leading to a strongly inhomogeneous distribution. At higher pre-strain values, however, a more uniform dislocation distribution was observed (Sudharshan Phani et al., 2011; Kwon et al., 2015). Although the inhomogeneous distribution as well as the impact of the interface are not yet fully understood, the strong increase of the dislocation density at small pre-strain values and the saturation at larger pre-strain values is clearly visible (compare Fig. 2.5). This behavior can be described by the dislocation evolution according to Mecking et al. (1976); Mecking and Kocks (1981); Kocks and Mecking (2003); Estrin (1996) and reads

$$\dot{\rho} = \left(k_1 \sqrt{\rho} - k_2 \rho\right) \dot{\gamma}_{\mathrm{acc}}. \tag{7.7}$$

Here, $\dot{\gamma}_{\mathrm{acc}} = \sum_{\alpha=1}^{N_{\mathrm{GS}}} |\dot{\gamma}_\alpha|$ corresponds to the total shear rate. This dislocation density evolution includes two physical dislocation interaction phenomena, the effective dislocation storage and the annihilation of dislocations described by the first and the second term, respectively. Calculating the dislocation density evolution with the help of Eq. (7.7) represents the discussed trend of the measurements with an initial strong increase as well as a final saturation (shown in Fig. 2.5). Here, the two parameters k_1 and k_2 are given in Tab. 7.1 and were identified by fitting to the experimentally measured values. The mean value of the two measured dislocation densities by Sudharshan Phani et al. (2011) at $4\,\%$ pre-strain were used for the identification of k_1 and k_2. Barabash et al. (2010a) estimated the dislocation density values, thus, these values were not incorporated.

The dislocation storage process is athermal and k_1 is of statistical nature, therefore, the identified parameter can be used at high temperatures (Mecking and Kocks, 1981), although the pre-strain was induced at room temperature. Contrary, the dislocation annihilation process is affected by thermal activation. Consequently, the determined dislocation annihilation parameter k_2 is useless at elevated temperatures. However, k_2 increases with increasing temperature, resulting in the condition

$$k_2^{\mathrm{RT}} < k_2^{1000\,^\circ\mathrm{C}}, \tag{7.8}$$

as stated in Mecking and Kocks (1981); Estrin (1996).

7.2.5 Simulation setup

The experimental microstructural investigations of several authors revealed:

1. The regular hexagonal arrangement of the fibers perpendicular to the growth direction (e.g., Bei and George (2007; 2005); Zhang et al. (2013); Haenschke et al. (2010)).
2. The perfect alignment of the fibers parallel to the growth direction (e.g., Barabash et al. (2010a); Bei et al. (2008a)).
3. The high aspect ratio of fiber length to fiber diameter (e.g., Hu et al. (2013); Haenschke et al. (2010); Bei et al. (2008a)).

Due to these findings, a periodic microstructure can be assumed (Albiez et al., 2016a), resulting in the possible use of a representative unit cell to describe the microstructure (discussed in Section 5.2). The microstructure of the DS NiAl-9Mo is depicted in the transverse section normal to the solidification direction (Fig. 7.2 a)) together with the representative unit cell (Fig. 7.2 b)). To benefit from the periodic microstructure, the finite element simulation was carried out by using the periodic displacement boundary conditions discussed in Section 5.3. Following the experimental results of, e.g., Dudová et al. (2011); Hu et al. (2012); Bei and George (2005); Hu et al. (2013), the $\langle 100 \rangle$-orientation of the fibers and the matrix were set parallel to the growth direction. Further, the cube-on-cube orientation relation of both phases with the interface boundary parallel to the $\{011\}$ crystal-plane (experimentally shown by Hu et al. (2012); Bei and George (2005); Hu et al. (2013)) was considered. True stress was applied and the strain was calculated as Hencky strain by using the displacement of a master node. Finally, due to the experimental challenge of identifying the precise zero point of plastic creep strain, the simulated minimum creep strain is set to the experimentally measured one.

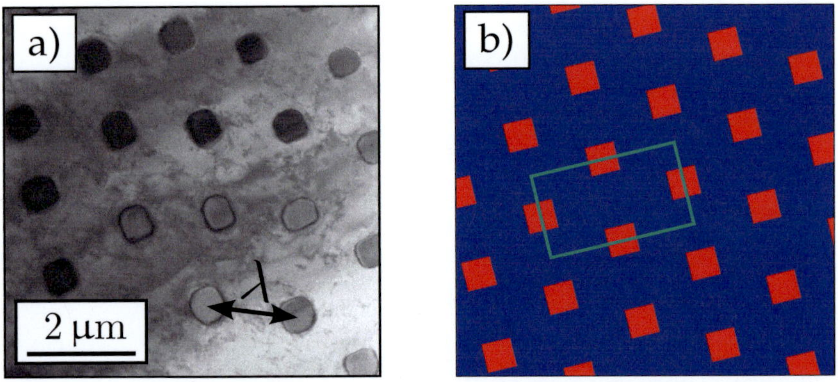

Figure 7.2: a) Transmission electron microscopy image showing the fibers embedded in the matrix (Albiez et al., 2016a). b) Geometrical discretization of the microstructure by assuming a perfect hexagonal fiber arrangement. The representative unit cell is marked with the rectangle. (Albiez et al. (2016a) Fig. 3)

7.3 Determination of material parameters

The previously introduced simulation model for the description of the creep behavior of DS NiAl-9Mo is based on measured quantities or material parameters restricted to a reasonable range of physically motivated values. For this reason, the required material parameters can be extracted from literature except for the reference shear rate of the fiber $\dot{\gamma}_0$ and the dislocation annihilation parameter k_2 at elevated temperatures, which were not present in literature. A stress exponent of $m = 10.5$ for the fiber material was chosen which is close to the value found in a one-dimensional model by Albiez et al. (2015). Further, it is close to the stress exponent of pure molybdenum (Pugh, 1955). The intrinsic size corresponds to the area-equivalent fiber diameter. To determine the two missing material parameters, namely $\dot{\gamma}_0$ and k_2, the simulation was fitted to one creep experiment from Albiez et al. (2015) at $1000\,^{\circ}\mathrm{C}$, an applied stress of $250\,\mathrm{MPa}$, and a fiber diameter of $0.729\,\mu\mathrm{m}$. The simulated creep curve is in good agreement with the experimental measured one, shown in Fig. 7.3 a). Tab. 7.1 summarizes the two identified parameters together with the necessary input parameters from literature. The identified dislocation annihilation parameter k_2 at $1000\,^{\circ}\mathrm{C}$ exceeds the value at RT, thus, fulfills the condition described in Eq. (7.8). Furthermore, due to the increased temperature, the dislocation densities are considerable lower than the ones measured at RT by Sudharshan Phani et al. (2011); Barabash et al. (2010a); Chisholm et al. (2012). Finally, as shown in Fig. 7.3 b), the fiber's flow stress is approximately half of the theoretical strength for $> 4\,\%$ creep strain.

Table 7.1: Identified and input parameters for the simulation of the DS NiAl-9Mo.

Description	Parameter	Value	Reference
Fiber			
Dislocation storage constant	k_1	$10^8\,\mathrm{m}^{-1}$	Fit Eq. (7.7) to ρ-measurement from: Sudharshan Phani et al. (2011), Barabash et al. (2010a), and Chisholm et al. (2012)
Dislocation annihilation constant	k_2	RT: 8	Fit Eq. (7.7) to ρ-measurement from: Sudharshan Phani et al. (2011), Barabash et al. (2010a), and Chisholm et al. (2012)
		900 °C and 1000 °C: 66.0	Albiez et al. (2016a)
Initial dislocation density	ρ_0	$9 \cdot 10^8\,\mathrm{m}^{-2}$	Sudharshan Phani et al. (2011)
Stress exponent	m	10.5	Pugh (1955); Albiez et al. (2015)
Activation energy (lattice diffusion)	$Q_\mathrm{c} = Q_\mathrm{SD}$	$405\,\mathrm{kJmol}^{-1}$	Frost and Ashby (1982)
Reference shear rate	$\dot{\gamma}_0$	$3.7 \cdot 10^{17}\,\mathrm{s}^{-1}$	Albiez et al. (2016a)
Theoretical strength	τ_∞	$\mu/30$	Cottrell (1953)
Constant	$\bar{\alpha}$	unity	Mecking and Kocks (1981)
Shear modulus	μ	27 °C: 134 GPa	Frost and Ashby (1982)
		900 °C: 116 GPa	Frost and Ashby (1982)
		1000 °C: 115 GPa	Frost and Ashby (1982)
Length of the Burgers vector of Mo	b	$2.73 \cdot 10^{-10}\,\mathrm{m}$	Frost and Ashby (1982)
Elastic constants (cubic symmetry)	$c_{1111}, c_{1122}, c_{1212}$	900 °C: 410, 163, 100 GPa	Wern (2004)
		1000 °C: 404, 163, 99 GPa	Wern (2004)
Fiber diameter	d	0.729 µm	Albiez et al. (2016a)
		0.784 µm	Albiez et al. (2016a)
		0.504 µm	Albiez et al. (2016a)
BCC glide systems	$\boldsymbol{n}_\alpha, \boldsymbol{d}_\alpha$	$\{110\}\langle 111\rangle$, $\{112\}\langle 111\rangle$, $\{123\}\langle 111\rangle$, $N_\mathrm{GS} = 48$	-
Schmid factor	s	0.47	Bei et al. (2007)
Matrix			
Description	Parameter	Value	Reference
Stress exponent	m	4.04	Whittenberger et al. (1991)
Activation energy	Q_c	$307.5\,\mathrm{kJmol}^{-1}$	Whittenberger et al. (1991)
Reference shear rate	$\dot{\gamma}_0$	$4.149 \cdot 10^9\,\mathrm{s}^{-1}$	Whittenberger et al. (1991)
Initial flow stress	τ_0	30.75 MPa	Whittenberger et al. (1991)
Elastic constants (cubic symmetry)	$c_{1111}, c_{1122}, c_{1212}$	900 °C: 184, 121, 88.1 GPa	Rusović and Warlimont (1977)
		1000 °C: 182, 120, 85.4 GPa	Rusović and Warlimont (1977)
B2 glide systems	$\boldsymbol{n}_\alpha, \boldsymbol{d}_\alpha$	$\{001\}\langle 100\rangle$, $\{011\}\langle 100\rangle$ $\{011\}\langle 110\rangle$, $N_\mathrm{GS} = 18$	Noebe et al. (1993)

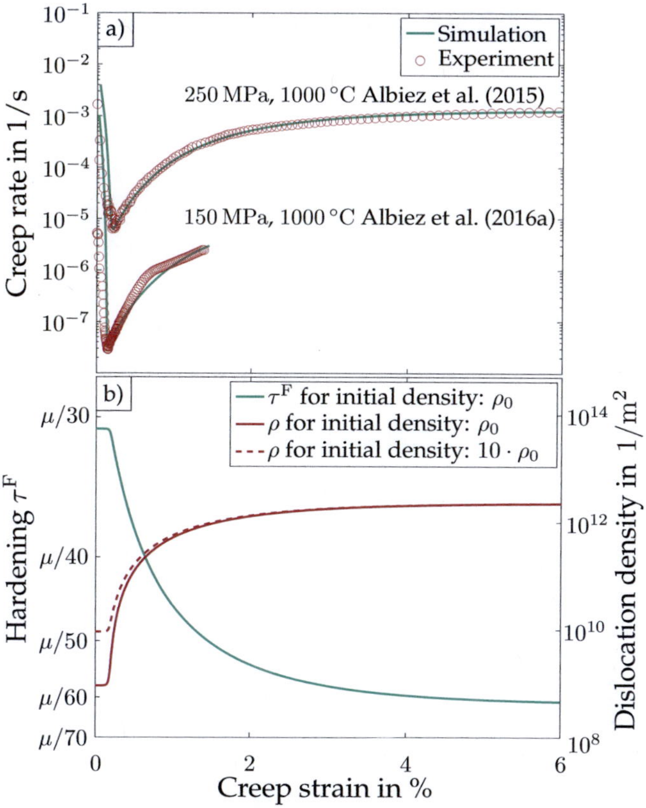

Figure 7.3: a) Comparison of the predicted creep curves with the experimentally measured ones for a fiber diameter of $d = 0.729\,\mu\mathrm{m}$ at $1000\,^\circ\mathrm{C}$ and an initial dislocation density of ρ_0. The experimentally measured creep curve at an applied stress of 250 MPa was published in Albiez et al. (2015) and was used to identify the two missing parameters ($\dot{\gamma}_0$ and k_2). The creep curve for an applied stress of 150 MPa is predicted and is in good agreement to the experiment (Albiez et al., 2016a). b) Evolution of the hardening τ^{F} (green curve) and the dislocation density ρ (red curves) of the fiber during creep deformation (at $1000\,^\circ\mathrm{C}$, an applied stress of 250 MPa, and a fiber diameter of $d = 0.729\,\mu\mathrm{m}$). An increase of the initial dislocation density by a factor of 10 only has a minor influence on the dislocation density evolution after reaching the minimum creep rate. (Albiez et al. (2016a) Fig. 4)

7.4 Evaluation of the creep model

7.4.1 Evaluation approach

To evaluate the simulation model proposed in Albiez et al. (2016a), the predicted creep curves were compared with experimentally identified ones. Only measured quantities or parameters which are restricted to a reasonable range of physically motivated values were used in this model. As emphasized earlier, a physically motivated model should be able to rationalize (i) the influence of a change of the applied stress, (ii) a modification of the temperature, and (iii) a change of the intrinsic size, thus, the fiber diameter. Further, the role of the inhomogeneous distribution of dislocations in the as-grown fibers on the minimum creep rate will be investigated. Finally, due to the use of the elastic constants of stoichiometric NiAl and pure Mo, the effects of the elastic constants on the creep behavior are considered.

7.4.2 Change of the applied stress

As discussed in Section 7.3, the reference shear rate of the fiber $\dot{\gamma}_0$ and the dislocation annihilation parameter k_2 were missing in literature, and, thus, were identified by fitting the simulation to one creep experiment published in Albiez et al. (2015). This deliberately picked creep experiment was conducted at $1000\,^{\circ}\mathrm{C}$, an applied stress of $250\,\mathrm{MPa}$, and a fiber diameter of $d = 0.729\,\mu\mathrm{m}$. To evaluate the predictions of the simulation model, the temperature as well as the fiber diameter are kept constant while the applied stress is decreased to $150\,\mathrm{MPa}$. The predicted creep curve and the experimentally measured one show a good agreement, as depicted in Fig. 7.3 a). Further, Fig. 7.4 shows the simulated minimum creep rate of the composite $\dot{\bar{\varepsilon}}_{\mathrm{min}}$ together with the experimentally one vs. the applied stress. Both, the predicted stress exponent and the experimentally measured one are around $\bar{m} \approx 10$, thus, match perfectly. Consequently, one can conclude that for the range of applied stresses covered by the experiments, the model reproduces well the effects of a change of the applied stress.

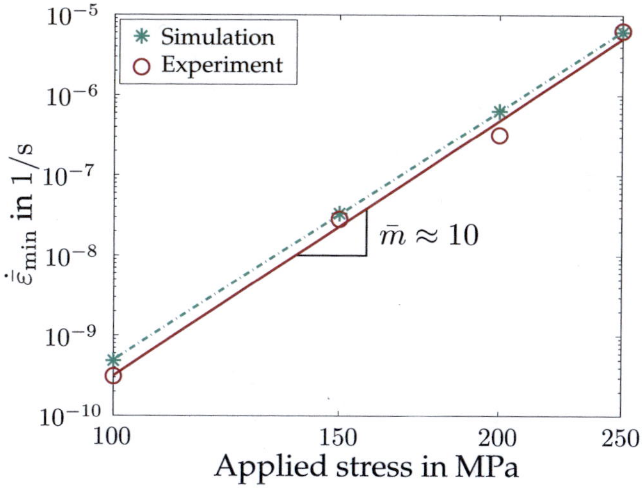

Figure 7.4: Experimentally measured minimum creep rates as well as simulated/predicted ones at 1000 °C and a fiber diameter of $d = 0.729\,\mu\text{m}$, represented in a Norton-plot. Both stress exponents are in perfect agreement. (Albiez et al. (2016a) Fig. 7)

7.4.3 Change of the temperature

The temperature has a strong impact on the slip rate as described by the Arrhenius term in Eq. (7.6). Consequently, evaluating the effects of a change in temperature to the response of the model is crucial. For this reason, the temperature was reduced from $1000\,°\text{C}$ to $900\,°\text{C}$, while keeping the applied stress of $250\,\text{MPa}$ and the fiber diameter of $d = 0.784\,\mu\text{m}$ close to the simulation of the parameter identification. The predicted creep curve at $900\,°\text{C}$ is shown in Fig. 7.5 b) together with the experimentally measured one. The minimum creep rate, an important characteristic value for a material under high temperature structural applications (Naumenko and Altenbach, 2016), can be well reproduced by the simulation model. The activation energy for creep of the composite was calculated as $\bar{Q}_{\mathrm{C}}^{\mathrm{Sim}} = 418\,\mathrm{kJ/mol}$, which is close to the value of Dudová et al. (2011), who measured an activation energy of $\bar{Q}_{\mathrm{C}}^{\mathrm{Exp}} = 401\,\mathrm{kJ/mol}$ experimentally. It is significantly higher than the value reported by Hu et al. (2013) ($\bar{Q}_{\mathrm{C}}^{\mathrm{Exp}} = 291 \pm 19\,\mathrm{kJ/mol}$), however, this value was measured by tension creep tests instead of compression creep tests. Even though the creep rate after exceeding the minimum is slightly underestimated, one can conclude that the predicted creep curve is in an acceptable agreement with the experimentally measured one. The difference between the simulation and the experiment in the tertiary regime can be explained by considering the dislocation annihilation parameter k_2. Although the temperature dependency of k_2 is known (e.g., Mecking and Kocks (1981); Kocks and Mecking (2003); Estrin (1996)), the value identified at $1000\,^{\circ}\mathrm{C}$ was also used for the prediction of the creep curve at $900\,^{\circ}\mathrm{C}$.

7.4.4 Change of the fiber diameter

The growth conditions strongly affect the fiber spacing and the fiber diameter of the DS NiAl-9Mo, where the latter lies in the range of $80\,\mathrm{nm} - 900\,\mathrm{nm}$ (discussed in Section 2.2.2). The results of creep simulations for specific fiber diameters are shown in Fig. 7.5. Here, the temperature of $T = 900\,^{\circ}\mathrm{C}$ and the applied stress of $250\,\mathrm{MPa}$ remain equal to the previously discussed predictions, thus, all changes can be attributed to the modification of the fiber diameter. To account for the fiber diameters covered by the experiments, it was reduced by a factor of ≈ 1.55 from $0.784\,\mathrm{\mu m}$ to $0.504\,\mathrm{\mu m}$. However, to consider the experimentally feasible range of withdrawal rates during directional solidification, two additional creep curves with a smaller diameter of $300\,\mathrm{nm}$ and a larger diameter of $1\,\mathrm{\mu m}$ are plotted in Fig. 7.5. This reveals the possible impact by a modification of the fiber diameter on the creep behavior. The minimum creep rate $\dot{\varepsilon}_{\mathrm{min}}$ remains unchanged with a reduction of the fiber diameter. Further, the decrease of the fiber diameter leads to a smoother increase of the creep rate after the minimum point. Both experimental observations can be well reproduced by the model. As emphasized in Section 7.4.3, the slightly underestimated increase of the predicted creep rate after the minimum point compared to the experimental creep rate can be attributed to the dislocation annihilation parameter k_2. In spite of this deviation, the model proposed in Albiez et al. (2016a) is applicable to various withdrawal rates during directional solidification of NiAl-9Mo.

7.4.5 Role of the initial dislocation density on the creep behavior

In order to account for the effects of the strongly inhomogeneous distribution of dislocations in the as-grown fibers, the initial dislocation density $\rho_0 = 9 \cdot 10^8\,\mathrm{m}^{-2}$ measured by Sudharshan Phani et al. (2011) was increased and decreased by a factor of 10. The simulated creep curves are shown in Fig. 7.6 and one observes a higher/lower minimum creep rate $\dot{\varepsilon}_{\mathrm{min}}$ due to a higher/lower initial dislocation density, respectively. The dashed line in Fig. 7.3 b) shows the corresponding dislocation evolution for an initial value of $10 \cdot \rho_0$, which approaches the dislocation evolution for an initial value of ρ_0 after exceeding the minimum creep rate. An increase of the initial dislocation density value from $0.1 \cdot \rho_0$ to $10 \cdot \rho_0$ leads to a change in the minimum creep rate from $5.4 \cdot 10^{-6}\,1/\mathrm{s}$ to $9.3 \cdot 10^{-6}\,1/\mathrm{s}$, consequently, the role of the initial dislocation density value on the minimum creep rate is small. This can be rationalized by considering the minor dependency of the hardening variable τ^{F} on the dislocation density for values below $10^{10}\,1/\mathrm{m}^2$ (shown in Fig. 6.1).

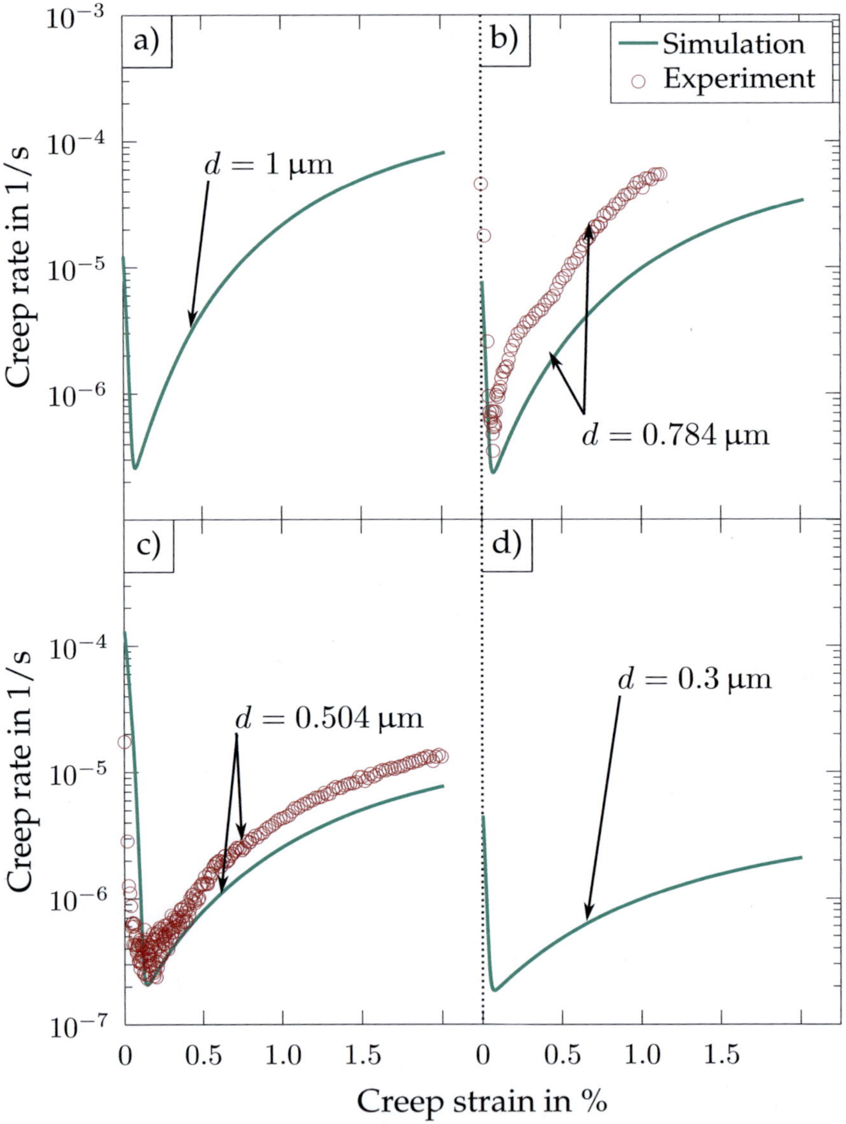

Figure 7.5: Predicted creep curves for different fiber diameters compared to experimentally measured ones at 900 °C and an applied stress of 250 MPa. (Albiez et al. (2016a) Fig. 6)

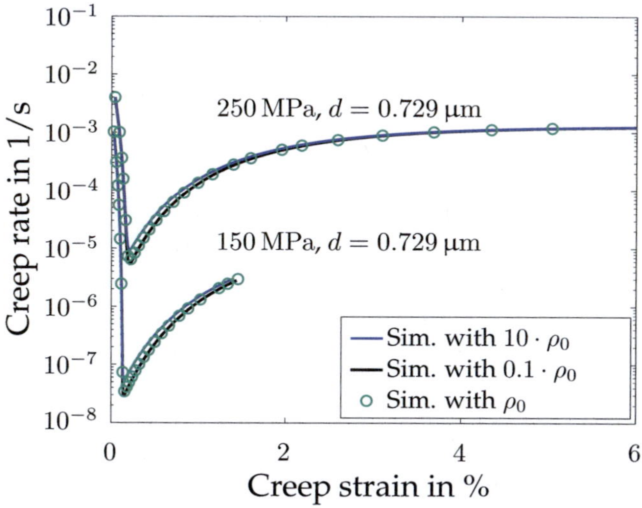

Figure 7.6: Analyzing the impact of the initial dislocation density at 1000 °C by increasing/decreasing the measured initial dislocation density value ρ_0 by a factor of 10. (Albiez et al. (2016a) Fig. 8)

Below this value, the hardening variable stays nearly constant, thus, a modification of the initial dislocation density ρ_0 has no significant effects on τ^{F}.

7.4.6 Role of the elastic constants on the creep behavior

For the elastic constants of the fibers and the matrix, the experimentally determined values of pure molybdenum (Wern, 2004) and stoichiometric NiAl (Rusović and Warlimont, 1977) were taken, respectively. The investigation of Bei and George (2007) reveals, however, that the chemical composition of the fibers is Mo-10Al-4Ni, whereas that of the matrix is Ni-45.2Al. This deviation of the chemical composition may affect the elastic properties. Therefore, creep curves were simulated for increased and decreased elastic constants of the fiber and the matrix, shown in Fig. 7.7. Obviously, an increase from $0.5 \cdot c_{ijkl}$ to $2 \cdot c_{ijkl}$, thus, an increase of the elastic constants by a factor of 4 only has a minor effect on the creep behavior.

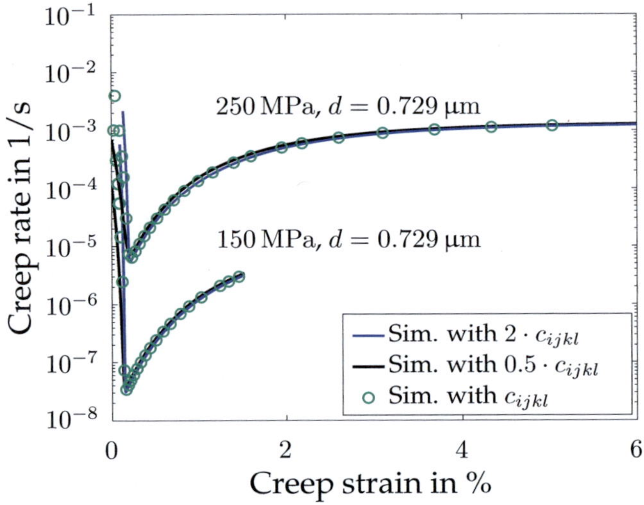

Figure 7.7: Analyzing the impact of the elastic constants on the creep behavior at 1000 °C.

7.5 Applicability of the creep model to directionally solidified eutectics with a lamellar microstructure

Directional solidification of the eutectic NiAl-(Cr,Mo) leads to a NiAl based matrix with (Cr,Mo) fibrous or lamellar reinforcements which are aligned parallel to the growth direction. As emphasized in Section 2.2.2, those DS eutectic composites exhibit an increased room temperature fracture toughness and an enhanced creep resistance simultaneously. The consideration of the transition from theoretical strength of the fibers to bulk strength was crucial in the case of the DS NiAl-9Mo, where the fibers are essentially dislocation free. As stated in Walter and Cline (1970), this is also the case for the DS NiAl-34Cr with fibrous reinforcements, therefore, it is likely that the lamellas within the DS NiAl-(Cr,Mo) are also essentially dislocation free. Consequently, the model originally proposed for the DS NiAl-9Mo might also be applicable for the DS NiAl-(Cr,Mo), even though one has a fibrous and one a lamellar microstructure. An initial attempt to apply this model also for lamellar materials was made in Albiez et al. (2016b). Here, the DS NiAl-31Cr-3Mo composite was used, which possesses a NiAl matrix reinforced by (Cr,Mo) lamellas. For the simulation, a periodic unit cell of perfectly aligned lamellas parallel to the load direction was used. Further, periodic displacement boundary conditions were applied. Consistent to the DS NiAl-9Mo case, an ideal plastic behavior $\tau^{\mathrm{F}} = \tau_0$ was used to describe the plastic deformation of the matrix. The hardening of the lamellas follows the transition law discussed in Chapter 6 (Eq. (6.3)), whereas the dislocation evolution was described by the well-known Kocks-Mecking formulation (Eq. (7.7)). The used material parameters are listed in Tab. 7.2. The simulated and the experimentally measured creep curves of the DS NiAl-31Cr-3Mo composite are shown in Fig. 7.8. The experiment was conducted by I. Sprenger and M. Heilmaier and published in Albiez et al. (2016b). Despite the difference between the simulation and the experiment in the primary creep regime, the minimum creep rate and the tertiary creep regime are in good agreement (Albiez et al., 2016b). Based on this acceptable agreement, one can assume a similar creep mechanism for the DS NiAl-31Cr-3Mo as observed for the DS NiAl-9Mo, even though the first has a lamellar reinforcing phase and the latter a fibrous reinforcing phase. The interplay between the NiAl matrix and the (Cr,Mo) lamellas can be explained as follows. In the initial stage, the (Cr,Mo) lamellas are essentially dislocation free, thus, the yield strength is close to the theoretical strength. Hence, the lamellas behave nearly linear elastic and the free plastic flow of the NiAl matrix results in a load transfer from the matrix to the lamellas. At the minimum creep rate, the stress in the lamellas reaches a critical value, which activates the slip of dislocations. Therefore, the dislocation density within the lamellas increases and thus, results in the softening of the lamellas (described by Eq. (6.3)). As a result of this softening, a reverse load transfer from the lamellas to the NiAl matrix occurs, which induces the increase of the creep rate in the tertiary creep regime.

Table 7.2: Identified and input parameters for the simulation of the DS NiAl-31Cr-3Mo.

Parameter	Value
	Lamella
k_1, k_2	$7 \cdot 10^7\,\mathrm{m}^{-1}$, 13
initial dislocation density	$10^9\,\mathrm{m}^{-2}$
m, $\dot{\gamma}_0$	$4.6, 0.4\,\mathrm{s}^{-1}$
$\bar{\alpha}$, μ, b	unity, 97.7 GPa, $2.5 \cdot 10^{-10}\,\mathrm{m}$ (Frost and Ashby, 1982)
d	$0.409\,\mu\mathrm{m}$
C_{1111}, C_{1122}, C_{1212}	350.0, 67.8, 100.8 GPa (Wern, 2004)
$\boldsymbol{n}_\alpha$, $\boldsymbol{d}_\alpha$	$\{110\}\langle 111\rangle$, $\{112\}\langle 111\rangle$, $\{123\}\langle 111\rangle$
	Matrix
m, $\dot{\gamma}_0$, τ_0	$5.75, 0.001\,\mathrm{s}^{-1}$, 37.25 MPa (Whittenberger, 1987)
c_{1111}, c_{1122}, c_{1212}	182, 120, 85.4 GPa (Wern, 2004)
$\boldsymbol{n}_\alpha$, $\boldsymbol{d}_\alpha$	$\{001\}\langle 100\rangle$, $\{011\}\langle 100\rangle$, $\{011\}\langle 110\rangle$

7.6 Discussion of the crystal plasticity creep model

The evaluation in Section 7.4 reveals the capability of the model to reproduce the influences of (i) a change of the applied stress, (ii) a modification of the temperature, and (iii) a change of the fiber diameter in the experimentally investigated range. To describe the softening of the fiber, a transition law proposed by Albiez et al. (2016a) as discussed in Chapter 6 was used. It was motivated by a regularization of the

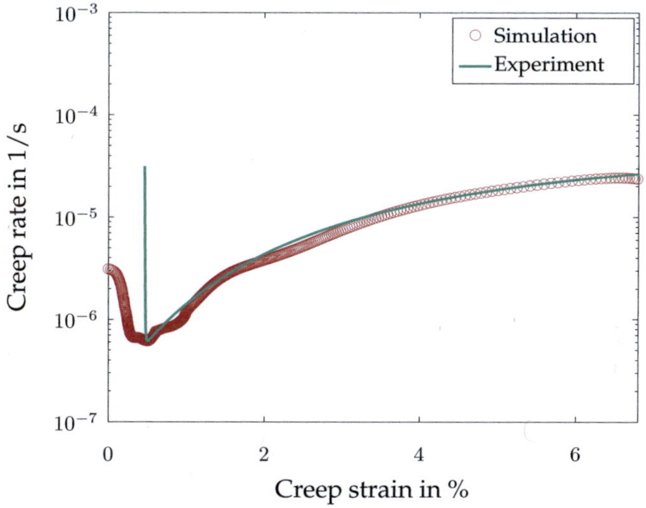

Figure 7.8: Comparison of the simulation with the experiment for a eutectic DS NiAl-31Cr-3Mo. The microstructure consists of a NiAl matrix reinforced by fine (Cr,Mo) lamellas. The results were obtained at 1000 °C and for an applied stress of 200 MPa. (Albiez et al. (2016b) Fig. 1)

hardening law proposed by El-Awady (2015). It is noteworthy that a combination of the transition law proposed by Albiez et al. (2016a) and a classical Arrhenius term for the temperature dependency can be used to describe the plastic behavior of the fiber in the case of elevated temperatures. The simulations and the experiments reveal a stress exponent $\bar{m} \approx 10$ and an activation energy for creep $\bar{Q}_{\mathrm{C}}^{\mathrm{Sim}} = 418\,\mathrm{kJ/mol}$ of the composite (Albiez et al., 2016a). These values are close to the values of pure molybdenum ($m^{\mathrm{Mo}} \approx 10.5$ and $Q_{\mathrm{C}}^{\mathrm{Mo}} = 405\,\mathrm{kJ/mol}$), consequently, it was concluded that the creep behavior of the composite is mainly defined by the plastic behavior of the fiber (Albiez et al., 2016a). Further, they are in agreement with the experimental results by Dudová et al. (2011). In contrast, the results by Hu et al. (2013) indicate an activation energy for creep close to the value of the NiAl matrix, however, those results were obtained by tension creep tests instead of compression creep tests, thus, are not directly comparable to the values of Dudová et al. (2011); Albiez et al. (2016a). A change of the fiber diameter d does not affect the minimum creep rate $\dot{\bar{\varepsilon}}_{\mathrm{min}}$ remarkably, exemplified in Fig. 7.5. This can be well explained by considering the low dislocation density within the fibers. The strength of a defect-free material is given by its theoretical strength ($\lim_{\rho\to 0}\tau^{\mathrm{F}} = \tau_{\infty}$), thus, it is independent of the intrinsic size. This behavior was experimentally shown for the fiber material by Bei et al. (2007); Sudharshan Phani et al. (2011); Chisholm et al. (2012); Kwon et al. (2015) and discussed in general in Chapter 6. Both, the dislocation multiplication parameter k_1 and the dislocation annihilation parameter k_2 were modeled independently of the fiber diameter. Consequently, the smoother increase of the creep rate with a decreasing fiber diameter can be referred to the transition law (Eq. (6.3)). Mathematically this can be rationalized by regarding the inequality $(\partial\dot{\gamma}_{\mathrm{acc}}/\partial\gamma_{\mathrm{acc}})|_{d=d_1} \geq (\partial\dot{\gamma}_{\mathrm{acc}}/\partial\gamma_{\mathrm{acc}})|_{d=d_2}$ which is fulfilled for $d_1 \geq d_2$, $\rho \leq 1/(d_1 d_2)$ and $\rho \leq (1/d_2^2)(\sqrt{(d_2)\tau_{\infty}/(\bar{\alpha}\mu b)} - 1)^2$. Hence, for the fiber material, the increase of the total shear rate $\dot{\gamma}_{\mathrm{acc}}$ with increasing shear γ_{acc} is larger with increasing fiber diameter. The shape of the composite's creep curve can be emphasized by analyzing the interplay between the matrix and the fiber. In the primary creep regime, the sharp reduction of the creep rate to the minimum point results from the load transfer from the matrix to the fiber. Here, the stress in the fiber increases with increasing creep strain as a result of the essential linear-elastic behavior. As a consequence of the essential linear-elastic behavior of the fiber, the dislocation density and therefore the hardening variable τ^{F} remain constant in the primary creep regime, shown in Fig. 7.3 b). In contrast, the stress in the matrix decreases due to its free plastic flow at elevated temperatures (experimentally shown by Bei et al. (2005)). It follows that the load is transferred from the matrix to the fiber within this regime. The load transfer was experimentally approved by the work of Dudová et al. (2011) as work hardening zones (WHZ) around the fiber. Investigations of the dislocations in the matrix by scanning transmission electron microscopy indicate the presence of $\langle 100 \rangle$-type dislocations in the NiAl matrix, which are swept towards the interphase boundary due to the deformation and, thus, creating the observed WHZ (Kwon et al., 2015). The effects of the pile-up stress at the interphase boundary cannot be considered within the simulation model, however, the load transfer leads to a strong increase of the stress in the fiber. At the minimum point, the stress reaches a critical value inducing the activation of slip, which leads to a sharp increase of the dislocation density in the fiber. This activation was experimentally observed by Kwon et al. (2015) as Hairpin-like dislocations which nucleated at stress concentrations at the fiber-matrix interface. The sharp increase of the dislocation density induces a strong softening of the fiber (depicted in Fig. 7.3 b)) and, as a consequence, an increase of the creep rate of the composite in the tertiary creep regime. The pronounced increase of the dislocation density results from the dislocation multiplication with creep strain, described by the first term in Eq. (7.7). Within the model, the slope of the creep curve right after the minimum point is controlled by the dislocation multiplication factor k_1 of Eq. (7.7) and the diameter d via the hardening τ^{F} (shown in Fig. 7.5). As emphasized in section 7.2.4 and following Mecking and Kocks (1981), the dislocation multiplication factor was assumed to be of purely statistical nature. Hence, it was assumed to be temperature independent and it was identified by fitting the experimentally obtained values at RT. Nevertheless, this is only true for the material's volumetric part, thus, the dislocation nucleation resulting at the fiber-matrix interface should be considered as an additional contribution. Further, this additional part should be modeled temperature dependent. The temperature dependent dislocation nucleation at the interface boundary can be motivated by Fig. 7.9 a), which shows the measured creep rates normalized by the corresponding minimum creep rate $\dot{\varepsilon}_{\min}$ over the ternary creep strain.

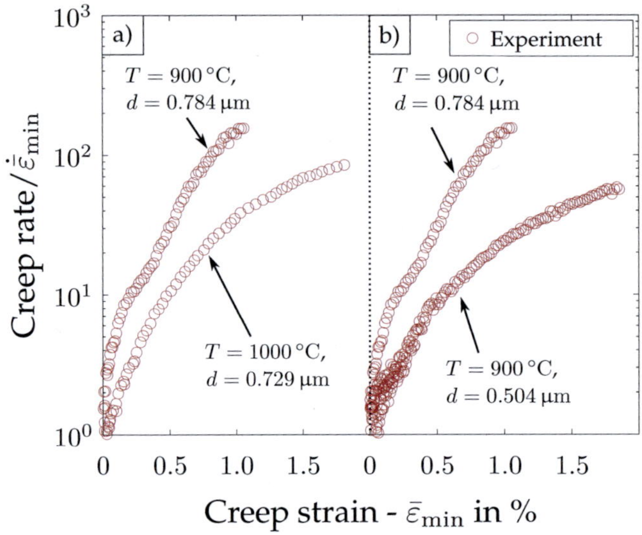

Figure 7.9: Creep rate normalized by the corresponding minimum creep rate $\dot{\bar{\varepsilon}}_{min}$ vs. ternary creep strain at an applied stress of 250 MPa. The slope of the curve directly after the minimum increases for a decreasing temperature and for an increasing fiber diameter. (Albiez et al. (2016a) Fig. 9)

The two curves depicted in Fig. 7.9 a) were measured for the same applied stress and almost the same fiber diameter, while changing the temperature. In a similar manner, Fig. 7.9 b) shows two curves, where the applied stress and the temperature remain equal, isolating the effect of a changing fiber diameter. The comparison of the slopes of the corresponding two creep curves right after the minimum in Fig. 7.9 a) and b) motivates that the dislocation multiplication should increase with decreasing temperature and with increasing fiber diameter (Albiez et al., 2016a). The saturation value of the dislocation density $\rho_{sat} = (k_1/k_2)^2$ is reached quickly and the strength of the fiber is still about half of the theoretical strength (shown in Fig. 7.3 b)). The transition law proposed in Albiez et al. (2016a) and described by Eq. (6.3) captures two strengthening mechanisms. To reveal the impact of those on the fiber strength, the saturated dislocation density can be used, which is about $\rho_{\text{sat}} \approx 2.3 \cdot 10^{12}\,1/\text{m}^2$ at $1000\,^\circ\text{C}$. Despite this maximal possible dislocation density, the dislocation-forest strengthening term in Eq. (6.3) only has a minor impact due to $\bar{\alpha}\mu b\sqrt{\rho_{\text{sat}}} \ll \tau_\infty/(\sqrt{\rho_{\text{sat}}}d + 1)$. In order to rationalize this finding, the mean dislocation spacing $1/\sqrt{\rho_{\text{sat}}}$ is compared with the intrinsic size d. Estrin (1996) stated that the fiber diameter corresponds to the smallest characteristic length if the inequality $d < 10/\sqrt{\rho_{\text{sat}}} \approx 6.6\,\mu\text{m}$ is fulfilled. Even at the saturated dislocation density, this inequality is clearly satisfied, thus, the fiber diameter can be identified as the smallest characteristic length. However, as the dislocation-forest strengthening mechanism is negligible, the nucleation and absorption of dislocations from the fiber-matrix interface should be considered in future studies. The second term in the Kocks-Mecking formulation (Eq. (7.7)) represents the mutual dislocation annihilation mechanism. The minimal mean dislocation spacing of $1/\sqrt{\rho_{\text{sat}}} \approx 0.66\,\mu\text{m}$ is, however, in the order of the fiber diameter, making the mutual dislocation annihilation questionable. Consequently, the dislocation annihilation due to the interface boundary should be considered. As proposed in Albiez et al. (2016a), this annihilation mechanism should scale as $1/d$, leading to a more likely dislocation annihilation for fibers with a smaller diameter. Fig. 7.10 shows the results of creep simulations, where the original dislocation density evolution (Eq. (7.7)) was replaced by

$$\dot{\rho} = \left(\tilde{k}_1\sqrt{\rho} - \frac{\tilde{k}_2}{d}\rho\right)\dot{\gamma}_{\text{acc}}. \tag{7.9}$$

Here, $\tilde{k}_1 = 1.6 \cdot 10^8\,\text{m}^{-1}$ and $\tilde{k}_2 = 51 \cdot 10^{-6}\,\text{m}$ were identified by fitting. The comparison between the simulation and the experiment for the two different fiber diameters shows a good agreement. For this reason one may conclude that the dislocation density annihilation scales by $1/d$. This behavior can be motivated by the interaction between the dislocations with the interface boundary and improves the simulation model of Albiez et al. (2016a).

The Kocks-Mecking formulation (Eq. (7.7)) does not account for static recovery. This can be rationalized by considering the duration for reaching the minimum creep rate for an applied stress of 150 MPa and 250 MPa at 1000 °C. The duration of the experiment for an applied stress of 150 MPa is more than 60 times longer compared to the duration for an applied stress of 250 MPa, however, the simulations agree perfectly with the experimentally measured creep curves. For this reason, the static recovery effects can be neglected. This was expected due to the low dislocation density in the fiber, where the same arguments can be used as for the mutual dislocation annihilation term (Albiez et al., 2016a). The mean spacing between the dislocations is in the order of d, therefore, the annihilation of two dislocations with opposite Burgers vectors is unlikely.

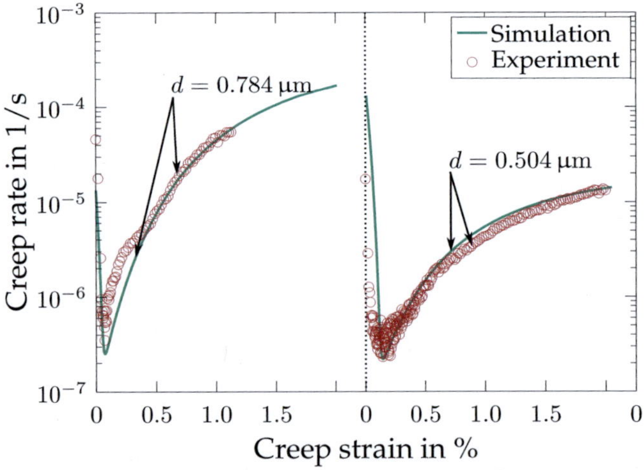

Figure 7.10: Results of the simulation model with a modified dislocation density evolution (Eq. (7.9)) are in good agreement with the experiments of Albiez et al. (2016a) (at 900 °C and an applied stress of 250 MPa). The adjusted dislocation evolution considers the $1/d$ scaling dislocation annihilation mechanism.

The chemical composition of the fibers is Mo-10Al-4Ni and that of the matrix Ni-45.2Al (Bei and George, 2007). For the simulation, however, the elastic constants of pure Mo and NiAl were used for the fiber and the matrix, respectively. This simplification is valid due to the minor effect of the elastic constants on the creep behavior (compare creep curves with increased and decreased elastic constants in Fig. 7.7). The minor effect can be rationalized by the small elastic strains and the low yield strengths within a creep test.

Investigations by Zhang et al. (2012) revealed the possibility to increase the fiber content while keeping the fibrous microstructure. They increased the volume fraction of the fiber up to approx. 17.6 vol.-% by the directional solidification of off-eutectic compositions. Fig. 7.11 shows the possible impact of the increased fiber content on the creep behavior of the composite by simulated creep curves. Here, the parameters identified in Section 7.4 and listed in Tab. 7.1 were used. For both applied loads, the minimum creep rate reduces by a factor of ≈ 10 with an increasing volume fraction from 14 vol.-% to 18 vol.-%. This behavior clearly shows the possible improvement of the directionally solidified material by using the approach proposed by Zhang et al. (2012). However, the off-eutectic composition may lead to a disturbed fiber alignment, thus, the increase of the creep resistance should be confirmed by experimental measurements, as the simulation assumes perfectly aligned fibers.

Finally, the simulation model was applied with a good agreement to the DS eutectic NiAl-31Cr-3Mo compound, even though this composite has a lamellar microstructure. Consequently, one can assume that the creep mechanism of both DS eutectics shows a similar behavior. Further, it is likely that the (Cr,Mo) lamellas are essentially dislocation free in the as-grown state.

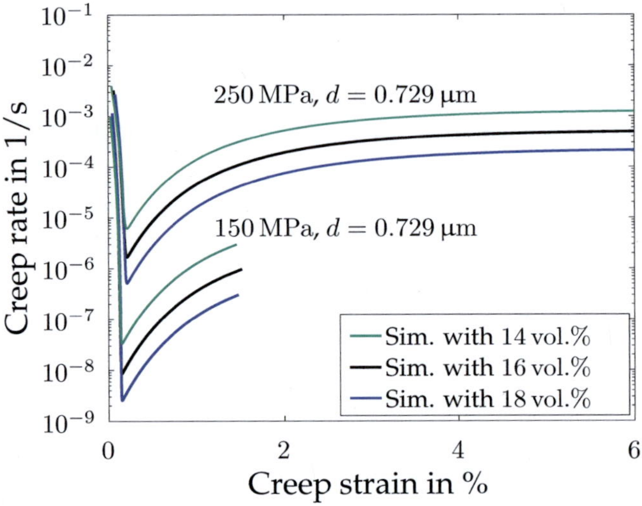

Figure 7.11: Simulated creep curves for different specific fiber contents at 1000 °C. An increasing fiber content leads to an improved creep resistance.

7.7 Summary and conclusion of the local simulation model

The new findings and insights due to the simulation model on the creep behavior of the DS NiAl based eutectic alloys will be summarized in this section. First, it is noteworthy that the model is based on measurable quantities or parameters restricted to a physically reasonable range. To evaluate the model, the experimentally measured creep curves were compared with predicted creep curves at different temperatures, applied stresses, and fiber diameters. The evaluation demonstrates the possibility to describe the creep behavior of DS NiAl-9Mo with the three-dimensional single-crystal plasticity model in the experimentally investigated range by Albiez et al. (2016a). The effects of (i) a change of the applied stress, (ii) a change of the temperature, and (iii) a change of the fiber diameter could be well described by the model. Further, it was applied to the DS NiAl-31Cr-3Mo with a lamellar microstructure with good agreement between the simulation and the experiment (Albiez et al., 2016b). The experimentally observed transition from the theoretical strength to the bulk strength of the fiber material was described by the transition law discussed in Chapter 6. With the help of the simulations, a closer insight into the state of the fibers and the matrix during the creep experiment is possible. One can conclude that the creep of the composite is controlled by the fibers, as the stress exponent and the activation energy for creep were in the order of the values of pure molybdenum. This is in line with the findings of Johnson et al. (1995). Moreover, the shape of the creep curve could be rationalized by considering the load transfer resulting from the interactions between the fibers and the matrix. Furthermore, the influences of the elastic properties on the creep curve were investigated. Finally, the impact of an increased fiber content on the creep behavior could be estimated. The simulations reveal that an increase of the volume fraction from $14\,\text{vol.-\%}$ to $18\,\text{vol.-\%}$ reduces the minimum creep rate by a factor of ≈ 10. However, the dislocation density evolution within the fiber, as depicted in Fig. 7.3 b), should be investigated experimentally to identify the dominant annihilation mechanism of dislocations for the fibers. In addition, the role of the interfaces on the creep behavior has to be further investigated.

Chapter 8

Gradient crystal plasticity model with interface yielding for creep of DS NiAl-9Mo

8.1 Introduction and motivation

The interplay between dislocations and microstructural constraints strongly affects the mechanical properties of single- and polycrystalline metals. As discussed in Section 3.3, the refinement of the grain size in polycrystals results in an increase of the yield strength, phenomenologically well described by the Hall-Petch relation (Hall, 1951; Petch, 1953). Despite this phenomenological theory, Zhu et al. (2008) concluded that many theories are still incomplete or only valid in a rather small domain of experimentally observed size effects. Consequently, a further improvement of the current theories is necessary. To explain the size effects in a physically motivated way, multiple gradient plasticity theories accounting for an internal length scale have been proposed (e.g., Fleck et al. (1994)). As stated by Ashby (1970), geometrically necessary dislocations (GNDs) are essential if two phases deform inhomogeneously. They affect the work hardening in two ways: as obstacles to hinder the movement of other dislocations and by causing long-range back stresses. To describe the GNDs, the Nye's dislocation density tensor (Nye, 1953) and the gradient of plastic slip can be used. Gurtin (2002) introduced a defect energy dependent on the Nye's dislocation density tensor, as an extension of the classical quadratic strain energy. Ohno and Okumura (2007) chose a defect energy based on the self-energy of GNDs and they were able to describe the d^{-1} grain size dependency of the initial yield strength. Svendsen and Bargmann (2010) discussed several GND models. Reddy et al. (2008) reformulated the gradient plasticity model proposed by Gurtin and Anand (2005a) as a variational inequality and showed the existence of a unique solution. The model of Gurtin and Anand (2005a) was valid for small deformations and later expanded for finite deformations in Gurtin and Anand (2005b). Specific possible defect energies including both, recoverable and non-recoverable terms are discussed in Reddy (2011a) for polycrystals and in Reddy (2011b) for single-crystals. A regularization of the logarithmic defect energy was introduced by Wulfinghoff et al. (2015), which captures the initial increase of the yield strength as well as the hardening effects. Anand et al. (2015) formulated a thermo-mechanically coupled gradient theory. However, higher order boundary conditions as a consequence of the extension of the gradient plasticity theories by an internal length scale are still necessary. The consideration of a microscopically powerless boundary condition leads to two important choices, namely the *microscopically free* and *microscopically hard* condition (Gurtin and Needleman, 2005). Energetically, the first one corresponds to the lower and the latter one to the upper limit case, respectively. No tractions of the gradient stresses (also called microstresses by several authors) at the boundary are implied in case of the *microscopically free* condition, consequently, the boundary does not hinder the dislocation flow. Contrary, dislocation transfer across the boundary is prohibited by the *microscopically hard* condition, thus, there is no plastic slip on the boundary. By considering a boundary energy, however, a behavior in between these two limit cases can be realized. Several approaches exist and could be used to explain experimentally observed size effects. With the help of the one-dimensional models described in Aifantis et al. (2006); Aifantis and Ngan (2007), the experimentally observed strain bursts during nanoindentation could be explained. Here, continuous slip across the grain boundaries with a quadratic boundary energy was considered. Further, to describe the Burgers vector production at the grain boundary, Gurtin and Needleman (2005) introduced the Burgers vector density of the grain boundary per unit length (also called grain boundary Burgers tensor). The rate of the Burgers vector density of the grain boundary corresponds to the Burgers vector production (e.g., Gurtin (2008)). Gurtin (2008) proposed a grain boundary energy which accounts for the grain misorientation and the boundary orientation. This framework, originally introduced for infinitesimal deformations, was later expanded by McBride et al. (2016) for finite deformations. In addition, Gottschalk et al. (2016) investigated the computational and theoretical aspects of this framework and reformulated it in a variational form. A rate and temperature dependent grain boundary flow rule was proposed in the work of Voyiadjis et al. (2014). The dependency of the dislocation nucleation and transmission on the stress state was investigated with several simulation methods. Bachurin et al. (2010); Tschopp et al. (2008) used atomistic simulations, De Koning et al. (2002) a combination of a molecular dynamic simulation with a line tension model, and Dewald and Curtin (2007) a coupled atomistic discrete-dislocation method. However, none of them can completely describe the currently observed strain rate sensitivity of slip transmission and the experimental observations (Malyar et al., 2017). Bayerschen et al. (2016) provide a review on the slip transmission criteria.

Several theoretical frameworks address the grain boundary behavior, however, comparisons between the simulation results and experimentally observed results are still challenging, especially in the three-dimensional case. This is also caused by the difficulties to extract the influences of the grain boundary from other effects. Therefore, a reliable and repeatable sample fabrication with a well-defined grain boundary structure, e.g., boundary orientation and grain misorientation is necessary. As discussed in Section 2.2.2, the directional solidification of the eutectic NiAl-9Mo offers such a well-defined phase boundary due to the following microstructural properties:

1. Adjustable fiber diameter in the range of approx. 100 nm (Bei et al., 2005) to approx. 900 nm (Zhang et al., 2013) with a constrained volume fraction of approx. 14 % due to the eutectic composition.
2. Well-defined cube-on-cube orientation relation between the fibers and the matrix, with the interface parallel to the $\{011\}$ crystal-plane (Bei and George, 2005; Hu et al., 2012; 2013).
3. Regular fiber arrangement with the $\langle 100 \rangle$-orientation of both phases lying parallel to the growth direction (Bei and George, 2005; Dudová et al., 2011; Hu et al., 2012; 2013).

The creep behavior of the composite could be well described by the simulation model proposed by Albiez et al. (2016a) and was discussed in Chapter 7. However, the effects of the fiber-matrix interface including potential dislocation transfer/nucleation have not been considered.

In this chapter, a three-dimensional gradient plasticity model with one accumulated plastic slip is used to account for the effects of dislocation pile-ups at the fiber-matrix interface. The simplification of accounting for only one accumulated plastic slip was originally proposed by Wulfinghoff and Böhlke (2012). Further, the activation of slip in the fibers as a result of the pile-up stress from the matrix dislocations is investigated. This was experimentally observed by Kwon et al. (2015). The dislocation transfer from the matrix into the fibers is modeled by a grain boundary yield condition. The evaluation of the grain boundary flow rule at the interfaces is performed by using the weak discontinuous shape functions introduced in Section 5.2.3. They are realized in the ABAQUS UEL finite element implementation. Even though, one cannot expect the same physical insights by using the simplification of considering only one accumulated plastic slip instead of taking all plastic slips into account individually, promising results have been received with this simplification (e.g., Wulfinghoff et al. (2013)). In addition, the numerical costs and the effort of considering all plastic slips individually exceed those of the simplified model several times.

8.2 Equilibrium conditions and constitutive relationships

8.2.1 Equilibrium conditions for the gradient plasticity theory

Contrary to the model introduced in the previous chapter, a micromorphic variable will be considered to account for the non-local effects of an inhomogeneous plastic deformation. The general procedure described in Section 4.5 can be used to derive the necessary equilibrium conditions, the jump conditions, and the boundary conditions of the gradient plasticity framework. Here, the micromorphic variable will be associated with the accumulated plastic slip γ_{acc}, thus, the general internal variable ϕ corresponds to γ_{acc} and the general micromorphic variable $\breve{\phi}$ to $\breve{\gamma}$. Further, the continuity of the micromorphic variable $\breve{\gamma}$ across the interface Γ is assumed (e.g., as in Wulfinghoff et al. (2013)), whereas this is not necessarily true for its gradient field ($\mathrm{grad}\,(\breve{\gamma})$). Due to the continuity of the micromorphic variable, the jump condition on Γ reduces to $\Xi = [\![\boldsymbol{\xi}]\!] \cdot \boldsymbol{n}_{\Gamma}$, where the general variable $\Xi^{\mathrm{s}}_{\mathrm{G}}$ was replaced by Ξ. All necessary conditions for the simplified gradient plasticity framework are summarized in Box 8.1. Here, the suffix $(\cdot)_{\mathrm{G}}$, indicating the general variables, was dropped.

Box 8.1: Equilibrium conditions, jump conditions, and boundary conditions for the gradient plasticity framework accounting for only one accumulated plastic slip.

Field variable	Equilibrium condition $\forall \boldsymbol{x} \in \mathcal{V}$	Jump condition on Γ	Boundary condition on $\mathcal{A}_\mathrm{t} \vee \mathcal{A}_\Xi$
$\boldsymbol{u}$	$\operatorname{div}(\boldsymbol{\sigma}) = \boldsymbol{0}$	$[\![\boldsymbol{\sigma}]\!]\boldsymbol{n}_\Gamma = \boldsymbol{0}$	$\boldsymbol{\sigma}\boldsymbol{n} = \bar{\boldsymbol{t}}$
$\breve{\gamma}$	$\pi = \operatorname{div}(\boldsymbol{\xi})$	$\Xi = [\![\boldsymbol{\xi}]\!] \cdot \boldsymbol{n}_\Gamma$	$\boldsymbol{\xi} \cdot \boldsymbol{n} = \bar{\Xi}$

8.2.2 Kinematical assumptions

In the gradient plasticity framework at hand, only infinitesimal deformations will be considered for sake of improved readability. Consequently, the simplifications discussed in Section 4.6 can be applied. The framework, however, can be easily expanded to finite deformations. The infinitesimal strain corresponds to the symmetric part of the displacement gradient tensor and is additively decomposed into an elastic part $\boldsymbol{\varepsilon}_\mathrm{e}$ as well as a plastic part $\boldsymbol{\varepsilon}_\mathrm{p}$, thus,

$$\boldsymbol{\varepsilon} = \operatorname{sym}(\boldsymbol{H}) = \boldsymbol{\varepsilon}_\mathrm{e} + \boldsymbol{\varepsilon}_\mathrm{p}. \tag{8.1}$$

The plastic part of the infinitesimal strain tensor reads

$$\boldsymbol{\varepsilon}_\mathrm{p} = \sum_{\alpha=1}^{N_\mathrm{GS}} \gamma_\alpha \boldsymbol{M}_\alpha^\mathrm{S} \tag{8.2}$$

and is defined by the slip in each slip system $\gamma_\alpha = \int \dot{\gamma}_\alpha \mathrm{d}t$. Further, $\boldsymbol{M}_\alpha^\mathrm{S} = \operatorname{sym}(\boldsymbol{d}_\alpha \otimes \boldsymbol{n}_\alpha)$ corresponds to the symmetric part of the Schmid tensor of the slip system α. The accumulated plastic slip is defined as $\gamma_\mathrm{acc} = \int \dot{\gamma}_\mathrm{acc} \mathrm{d}t$, whereas the rate of the accumulated plastic slip accounts for the amount of the rates of plastic slip for all slip systems, thus,

$$\dot{\gamma}_\mathrm{acc} = \sum_{\alpha=1}^{N_\mathrm{GS}} |\dot{\gamma}_\alpha|. \tag{8.3}$$

The number of slip systems N_{GS} depends on the material and is listed in Tab. 8.1.

8.2.3 Stored energy density for the matrix and the fibers

In this section, the constitutive assumptions for the matrix as well as for the fibers will be derived in a thermodynamically consistent framework. The free energy density in the bulk $W_{\mathcal{V}}$ includes an elastic W_{e}, a defect W_{∇}, and a penalty W_{χ} energy contribution. As stated in Ortiz and Repetto (1999), an additive decomposition between the elastic and the plastic contributions of the free energy is a good first approximation for metals. Therefore, following this suggestion, the free energy density in the bulk is given by

$$W_{\mathcal{V}} = W_{\mathrm{e}} + W_{\nabla} + W_{\chi}. \tag{8.4}$$

Here, contrary to the work of, e.g., Wulfinghoff et al. (2013), no hardening energy contribution is considered. This can be rationalized by considering the high temperatures during creep tests, where local hardening effects may be dissipated in time. Therefore, they do not contribute to the stored energy of the bulk. In addition, despite the possible description of the classical isotropic hardening in an energetical way, it can also be described in a dissipative way (Gurtin and Reddy, 2009). The elastic contribution to the free energy density is given by

$$W_{\mathrm{e}} = \frac{1}{2}\boldsymbol{\varepsilon}_{\mathrm{e}} \cdot \mathbb{C}[\boldsymbol{\varepsilon}_{\mathrm{e}}]. \tag{8.5}$$

Here, $\mathbb{C}$ corresponds to the elastic stiffness tensor with the elastic constants summarized in Tab. 8.1 for the matrix and the fibers. Several defect energies exist to include the internal length scale (e.g., Bayerschen and Böhlke (2016); Reddy (2011b)), however, in the gradient plasticity model at hand, a quadratic defect energy

$$W_{\nabla} = \frac{1}{2}k_{\nabla}\mathrm{grad}\,(\check{\gamma}) \cdot \mathrm{grad}\,(\check{\gamma}) \tag{8.6}$$

is chosen with a constant defect parameter k_∇. This defect energy leads to a gradient stress $\boldsymbol{\xi}$ linear in the gradient $\mathrm{grad}\,(\check{\gamma})$. Furthermore, the gradient $\mathrm{grad}\,(\check{\gamma})$ can be understood as an approximation for the geometrically necessary dislocation density, consequently, a quadratic defect energy leads to a gradient stress $\boldsymbol{\xi}$ linear in the geometrically necessary dislocation density (Wulfinghoff et al., 2013). As discussed in Section 4.5, the coupling between the micromorphic variable $\check{\gamma}$ and its counterpart γ_{acc} is realized by a penalty energy

$$W_\chi = \frac{H_\chi}{2}(\gamma_{\mathrm{acc}} - \check{\gamma})^2, \tag{8.7}$$

where the penalty parameter H_χ has to be chosen sufficiently large. Resulting from the existence of the interface, an energy density per unit area contribution on the interface W_Γ is considered. This contribution is assumed to depend linearly on $\check{\gamma}$, thus,

$$W_\Gamma = k_\Gamma \check{\gamma}, \tag{8.8}$$

where the material parameter k_Γ corresponds to the constant resistance of the interface against slip transfer. Consequently, no hardening of the interface is considered (in contrast to, e.g., Bayerschen et al. (2016)). This can be rationalized by the same arguments as for the bulk free energy density. In the high temperature case, possible hardening effects of the interface can dissipate in time. Further, this linear dependency is consistent to previous investigations (e.g., Wulfinghoff et al. (2013)).

8.2.4 Dissipation inequality

By reformulating the general expression of the power of internal forces (Eq. (4.35)) by replacing the general variables with the corresponding variables of the gradient plasticity framework at hand, the power of internal forces follows

$$\mathcal{P}_{\text{int}} = \int\limits_{\mathcal{V}} \left(\boldsymbol{\sigma} \cdot \dot{\boldsymbol{\varepsilon}} + \boldsymbol{\xi} \cdot \text{grad}\left(\dot{\check{\gamma}}\right) + \pi\dot{\check{\gamma}}\right) \text{d}v + \int\limits_{\Gamma} \Xi\dot{\check{\gamma}}\text{d}a. \tag{8.9}$$

Here, the continuity of $\check{\gamma}$ across the interface Γ as well as the simplifications of the infinitesimal deformation framework were used. The internal power in the bulk material with the Cauchy stress $\boldsymbol{\sigma}$ conjugated to $\dot{\boldsymbol{\varepsilon}}$, the micromorphic stress $\boldsymbol{\xi}$ conjugated to $\text{grad}\left(\dot{\check{\gamma}}\right)$, and the micromorphic stress π conjugated to $\dot{\check{\gamma}}$ is given by the first integral term. Further, the second integral term describes the interface with a micromorphic traction Ξ conjugated to $\dot{\check{\gamma}}$.
The total dissipation $\mathcal{D}_{\text{tot}}$ reads in the isothermal case

$$\begin{aligned} \mathcal{D}_{\text{tot}} &= \mathcal{P}_{\text{ext}} - \int\limits_{\mathcal{V}} \dot{W}_{\mathcal{V}}\text{d}v - \int\limits_{\Gamma} \dot{W}_{\Gamma}\text{d}a \\ &= \int\limits_{\mathcal{V}} \boldsymbol{\sigma} \cdot \dot{\boldsymbol{\varepsilon}} + \boldsymbol{\xi} \cdot \text{grad}\left(\dot{\check{\gamma}}\right) + \pi\dot{\check{\gamma}}\text{d}v \\ &+ \int\limits_{\Gamma} \Xi\dot{\check{\gamma}}\text{d}a - \int\limits_{\mathcal{V}} \dot{W}_{\mathcal{V}}\text{d}v - \int\limits_{\Gamma} \dot{W}_{\Gamma}\text{d}a \geq 0. \end{aligned} \tag{8.10}$$

Here, the power of external forces $\mathcal{P}_{\text{ext}}$ was expressed by the power of internal forces $\mathcal{P}_{\text{int}}$ due to the relation $\mathcal{P}_{\text{ext}} = \mathcal{P}_{\text{int}}$. First, the dissipation in the bulk material $\mathcal{D}_{\mathcal{V}}$ will be addressed. Substitution of the stored energy density of the bulk (Eq. (8.4)) into the dissipation inequality (Eq. (8.10)) and applying the material time derivative leads to

$$\begin{aligned} \mathcal{D}_{\mathcal{V}} = \int\limits_{\mathcal{V}} &\left(\boldsymbol{\sigma} - \frac{\partial W_{\text{e}}}{\partial \boldsymbol{\varepsilon}}\right) \cdot \dot{\boldsymbol{\varepsilon}} + \left(\boldsymbol{\xi} - \frac{\partial W_{\nabla}}{\partial \text{grad}\left(\check{\gamma}\right)}\right) \cdot \text{grad}\left(\dot{\check{\gamma}}\right) \\ &+ \left(\pi - \frac{\partial W_{\chi}}{\partial \check{\gamma}}\right)\dot{\check{\gamma}} - \left(\frac{\partial W_{\text{e}}}{\partial \boldsymbol{\varepsilon}_{\text{p}}} \cdot \dot{\boldsymbol{\varepsilon}}_{\text{p}} + \frac{\partial W_{\chi}}{\partial \gamma_{\text{acc}}}\dot{\gamma}_{\text{acc}}\right) \text{d}v \geq 0. \end{aligned} \tag{8.11}$$

The stresses $\boldsymbol{\sigma}$, $\boldsymbol{\xi}$, and π are considered to be purely energetic. Furthermore, by assuming arbitrariness of the rates $\dot{\boldsymbol{\varepsilon}}$, $\dot{\check{\gamma}}$, and $\text{grad}\left(\dot{\check{\gamma}}\right)$, the following relations are obtained:

$$\boldsymbol{\sigma} = \frac{\partial W_{\mathrm{e}}}{\partial \boldsymbol{\varepsilon}} = \mathbb{C}[\boldsymbol{\varepsilon}_{\mathrm{e}}], \qquad \check{p} := \frac{\partial W_{\chi}}{\partial \gamma_{\mathrm{acc}}} = -\frac{\partial W_{\chi}}{\partial \check{\gamma}} = H_{\chi}\left(\gamma_{\mathrm{acc}} - \check{\gamma}\right),$$

$$\boldsymbol{\xi} = \frac{\partial W_{\nabla}}{\partial \mathrm{grad}\left(\check{\gamma}\right)} = k_{\nabla} \mathrm{grad}\left(\check{\gamma}\right), \quad \pi = \frac{\partial W_{\chi}}{\partial \check{\gamma}} = -\check{p}. \tag{8.12}$$

Here, the abbreviation $\check{p}$ as the generalized stress is introduced. By using the previous relations (Eq. (8.12)), the dissipation inequality of the bulk material $\mathcal{D}_{\mathcal{V}}$ reduces to

$$\mathcal{D}_{\mathcal{V}} = \int_{\mathcal{V}} \boldsymbol{\sigma} \cdot \dot{\boldsymbol{\varepsilon}}_{\mathrm{p}} - \check{p}\dot{\gamma}_{\mathrm{acc}} \mathrm{d}v = \int_{\mathcal{V}} \sum_{\alpha=1}^{N_{\mathrm{GS}}} (\mathrm{sgn}\left(\dot{\gamma}_{\alpha}\right) \tau_{\alpha} - \check{p}) |\dot{\gamma}_{\alpha}| \mathrm{d}v \geq 0. \tag{8.13}$$

Here, in addition to equations (3.4), (8.2), and (8.3), the relations

$$\frac{\partial W_{\mathrm{e}}}{\partial \boldsymbol{\varepsilon}_{\mathrm{p}}} = -\boldsymbol{\sigma}, \tag{8.14}$$

$$\frac{\partial W_{\chi}}{\partial \gamma_{\mathrm{acc}}} = \check{p} \tag{8.15}$$

were used. The reduced dissipation inequality can help to motivate a thermodynamically consistent flow rule. This will be addressed in the following section (Section 8.2.5).

By considering the interface terms of the total dissipation (Eq. (8.10)), the dissipation of the boundary interface $\mathcal{D}_{\Gamma}$ can be obtained

$$\begin{aligned} \mathcal{D}_{\Gamma} &= \int_{\Gamma} \left(\Xi - \frac{\partial W_{\Gamma}}{\partial \check{\gamma}}\right) \dot{\check{\gamma}} \mathrm{d}a \\ &= \int_{\Gamma} \left(\Xi - \Xi^{\mathrm{en}}\right) \dot{\check{\gamma}} \mathrm{d}a = \int_{\Gamma} \Xi^{\mathrm{dis}} \dot{\check{\gamma}} \mathrm{d}a \geq 0. \end{aligned} \tag{8.16}$$

Here, analog to the approach for the bulk dissipation, the material time derivative of the stored energy density of the boundary interface (Eq. (8.4)) was substituted into the dissipation inequality (Eq. (8.10)).

Furthermore, the micromorphic traction Ξ was additively decomposed into an energetic part Ξ^{en} and a dissipative part Ξ^{dis}. It follows for the energetic part

$$\Xi^{\mathrm{en}} = \frac{\partial W_\Gamma}{\partial \check{\gamma}} = k_\Gamma \tag{8.17}$$

due to the linear dependency of W_Γ on $\check{\gamma}$ (compare Eq. (8.8)).

8.2.5 Flow rules

A thermodynamically consistent flow rule for the bulk material is given by

$$\dot{\gamma}_\alpha = \dot{\gamma}_0 \mathrm{sgn}\left(\tau_\alpha\right) \left(\frac{\langle |\tau_\alpha| - \check{p}\rangle}{\tau_\alpha^{\mathrm{F}}}\right)^m \tag{8.18}$$

and can be motivated by regarding the reduced dissipation inequality (Eq. (8.13)). Similar to the local model discussed in Chapter 7, the stress exponent, the reference shear rate, and the hardening variable in each slip system are denoted by m, $\dot{\gamma}_0$, and τ_α^{F}, respectively. The simplification of an isotropic hardening approach $\tau_\alpha^{\mathrm{F}} = \tau^{\mathrm{F}}$ is again used (rationalized in Section 7.2.2). The long-range back stresses, induced by the GNDs, will be considered in the flow behavior (Eq. (8.18)) by $\check{p}$ and the effects of the SSDs are included by τ^{F}. Therefore, both impacts are accounted for seperately. In addition, the simplification of accounting for only one accumulated plastic slip leads to a unique $\check{p}$ stress value, thus, it is equal on all slip systems.

The yield condition for plastic slip at the boundary interface

$$0 = \left(\Xi^{\mathrm{dis}} - \Xi^{\mathrm{F}}\right)\dot{\check{\gamma}} = \underbrace{\left(\Xi - k_\Gamma - \Xi^{\mathrm{F}}\right)}_{=:\, f_\Gamma}\dot{\check{\gamma}} = f_\Gamma \dot{\check{\gamma}} \tag{8.19}$$

can be motivated by considering the boundary interface dissipation inequality (Eq. (8.16)). Here, f_Γ as an abbreviation and $\Xi^{\mathrm{F}} \geq 0$ as a dissipative contribution to the yield strength were introduced. Further, the Kuhn-Tucker conditions for a rate-independent interface behavior are given by

$$f_\Gamma \leq 0, \quad \dot{\breve{\gamma}} \geq 0, \quad \dot{\breve{\gamma}} f_\Gamma = 0. \tag{8.20}$$

Several investigations focused on the dependency of the dislocation nucleation and the slip transfer at the boundary on the stress state. De Koning et al. (2002) used a combination of a molecular dynamic simulation with a line tension model. They revealed the impact of the ratio of resolved stress on the incoming to that on the outgoing slip system on the nucleation of dislocations at the grain boundary. In addition, Tschopp et al. (2008) investigated the nucleation process with the help of an atomistic simulation and concluded that the normal stress on the nucleation slip system influences the nucleation process. Finally, by using a coupled atomistic discrete-dislocation method, Dewald and Curtin (2007) stated that nucleation of dislocations is affected by the normal stress on the grain boundary. The exact effect of the stress state on dislocation nucleation and slip transfer at the boundary interface is still unclear, however, stresses other than the residual shear stress should be considered (Bachurin et al., 2010). Following the suggestion of Dewald and Curtin (2007), a dissipative contribution to the yield strength of the interface accounting for the normal stress on the interface

$$\Xi^{\mathrm{F}} = k_\sigma \langle -\boldsymbol{\sigma} \cdot (\boldsymbol{n}_\Gamma \otimes \boldsymbol{n}_\Gamma) \rangle \tag{8.21}$$

is considered. Here, due to the Macaulay bracket, only a compressive stress on the interface is taken into account and scales with the parameter k_σ. This compressive stress dependency is in agreement with the results of Dewald and Curtin (2007), who revealed a strong increase of the necessary nucleation stresses due to the pressure at the interface. Further, it can be rationalized by considering the volume of a disturbed crystal structure at the interface. This volume is relevant for the dislocation nucleation process. Therefore, a compressive stress on the interface hinders the nucleation process due to the reduction of this volume. In conclusion, the thermodynamically consistent flow rules were derived for the bulk material (Eq. (8.18)) as well as for the interface (Eq. (8.19)) by regarding the dissipation inequality.

8.2.6 Hardening model of the matrix

As emphasized in Section 7.2.3, hardening within the NiAl matrix can be neglected. Therefore, the plastic deformation of the matrix is modeled by an ideal plastic behavior, thus, $\tau^{\mathrm{F}} = \tau_0$. However, accumulation of GNDs in the matrix at the interface due to an inhomogeneous plastic deformation was experimentally observed after creep tests (e.g., Dudová et al. (2011)). Consequently, the effects of the GNDs in the matrix phase, inducing long-range back stresses, are considered by $\check{p}$ in the flow rule (Eq. (8.18)).

8.2.7 Hardening model of the fibers

To describe the yield strength of the fibers, a transition law from the theoretical strength to bulk strength is necessary. A physically motivated transition law is given by equation (6.3) and was introduced in Chapter 6. It describes the transition of the flow stress between different strengthening mechanisms and suggests that the flow stress should decrease starting from the theoretical strength for low dislocation densities and should reach the Taylor hardening for high dislocation densities. Good agreement between predicted creep curves and experimentally measured ones was achieved by using this transition law to describe the fibers' behavior during creep simulations (shown in Albiez et al. (2016a) and discussed in Chapter 7).

In the motivation of the transition model, Albiez et al. (2016a) considered the total dislocation density ρ and did not distinguish between the statistically stored dislocations ρ_{ssd} and the geometrically necessary dislocations ρ_{GND}. In the gradient plasticity framework at hand, however, the impact of the GNDs due to the long-range back stresses are accounted for in the flow rule (Eq. (8.18)) by $\check{p}$. Consequently, the local hardening model (Eq. (6.3)) will only consider the statistically stored dislocations, thus,

$$\tau^{\mathrm{F}} = \frac{\tau_\infty}{d\sqrt{\rho_{\mathrm{ssd}}} + 1} + \bar{\alpha}\mu b\sqrt{\rho_{\mathrm{ssd}}}. \tag{8.22}$$

It follows that due to this partition, the impacts of the statistically stored dislocations and the geometrically necessary ones are considered seperately through τ^{F} and $\check{p}$, respectively.

8.2.8 Dislocation evolution in the fibers

The development of the dislocation density within the fibers is described by an evolution law. As discussed in Section 2.2.2 and depicted in Fig. 2.5, the experimental investigations show that (i) the as-grown fibers essentially possess no defects (Sudharshan Phani et al., 2011) and (ii) the dislocation density saturates at higher pre-strains (Barabash et al., 2010a; Chisholm et al., 2012; Sudharshan Phani et al., 2011). During sample preparation for dislocation density measurements, the matrix was etched away, thus, the geometrical constraints on the fibers were eliminated (Barabash et al., 2010a). Therefore, the accumulated pile-up dislocations at the interface during pre-straining can escape through the free surface after this etch process. Consequently, the measured dislocation densities can be interpreted as SSDs (Albiez et al., 2018), even though the distinction between SSDs and GNDs is not possible within the dislocation density measurements.
The dislocation evolution was described in Albiez et al. (2016a) by the well-known Kocks-Mecking dislocation evolution (Eq. (7.7)) according to Estrin (1996); Kocks and Mecking (2003); Mecking and Kocks (1981); Mecking et al. (1976). As discussed in detail in Section 7.2.4, this dislocation evolution captures the effective dislocation storage

by $k_1\sqrt{\rho_{\text{ssd}}}$ and the annihilation of dislocations by $k_2\rho_{\text{ssd}}$. Therefore, it describes two physical dislocation interaction phenomena. Both necessary parameters, k_1 and k_2, were identified by Albiez et al. (2016a). However, they analyzed the validity of the k_2 parameter and concluded that the dislocation annihilation due to the fiber-matrix interface is more likely than the mutual dislocation annihilation mechanism. For this reason, and also to keep the model simple, the dislocation annihilation is neglected, thus $k_2 = 0$. In the infinitesimal deformation framework at hand, this assumption is reasonable, since the mutual dislocation annihilation is only relevant for higher dislocation densities. Those high density values occur only at high plastic slips, which cannot be investigated within the infinitesimal deformation framework. Nevertheless, the value for the dislocation multiplication (k_1), which was identified through fitting by Albiez et al. (2016a), can be used in the current gradient plasticity model. Consequently, the dislocation evolution

$$\dot{\rho}_{\text{ssd}} = k_1\sqrt{\rho_{\text{ssd}}}\dot{\check{\gamma}} \tag{8.23}$$

is used in this model with the k_1 parameter listed in Tab. 8.1.

Table 8.1: Identified and input parameters from literature for the gradient crystal plasticity model of the DS NiAl-9Mo composite.

Description	Parameter	Value	Reference
		Fiber	
Dislocation storage constant	k_1	$10^8\,\mathrm{m}^{-1}$	Albiez et al. (2016a)
Initial dislocation density	ρ_0	$9 \cdot 10^8\,\mathrm{m}^{-2}$	Sudharshan Phani et al. (2011)
Stress exponent	m	10.5	Albiez et al. (2015); Pugh (1955)
Reference shear rate	$\dot{\gamma}_0$	$78\,\mathrm{s}^{-1}$	This study
Theoretical strength	τ_∞	$\mu/30$	Cottrell (1953)
Constant	$\bar{\alpha}$	unity	Mecking and Kocks (1981)
Shear modulus	μ	$27\,°\mathrm{C}$: $134\,\mathrm{GPa}$, $1000\,°\mathrm{C}$: $115\,\mathrm{GPa}$	Frost and Ashby (1982)
Burgers vector length of Mo	b	$2.73 \cdot 10^{-10}\,\mathrm{m}$	Frost and Ashby (1982)
Elastic constants (cubic symmetry)	c_{1111}, c_{1122}, c_{1212}	$1000\,°\mathrm{C}$: $362\,\mathrm{GPa}$, $151\,\mathrm{GPa}$, $94\,\mathrm{GPa}$	Alers et al. (1960); Wern (2004); Gerlich and Fisher (1969)
Fiber diameter	d	$0.729\,\mu\mathrm{m}$	Albiez et al. (2016a)
Lamellar spacing	λ	$2\,\mu\mathrm{m}$	Albiez et al. (2016a)
BCC glide systems	$\boldsymbol{n}_\alpha$, $\boldsymbol{d}_\alpha$	$\{110\}\langle 111\rangle$, $N_{\mathrm{GS}} = 12$	-
Defect energy parameter	$k_\nabla = k_0\bar{\alpha}\mu b$	$27\,°\mathrm{C}$: $\approx 35 \cdot 10^{-6}\,\mathrm{N}$ with $k_0 = 1\,\mu\mathrm{m}$	-
		Matrix	
Stress exponent	m	4.04	Whittenberger et al. (1991)
Reference shear rate	$\dot{\gamma}_0$	$10^{-3}\,\mathrm{s}^{-1}$	Whittenberger et al. (1991)
Initial flow stress	τ_0	$30.75\,\mathrm{MPa}$	Whittenberger et al. (1991)
Elastic constants (cubic symmetry)	c_{1111}, c_{1122}, c_{1212}	$1000\,°\mathrm{C}$: $182\,\mathrm{GPa}$, $120\,\mathrm{GPa}$, $85.4\,\mathrm{GPa}$	Rusović and Warlimont (1977)
B2 glide systems	$\boldsymbol{n}_\alpha$, $\boldsymbol{d}_\alpha$,	$\{011\}\langle 110\rangle$, $N_{\mathrm{GS}} = 6$	Noebe et al. (1993)
Shear modulus	μ	$25\,°\mathrm{C}$: $71.5\,\mathrm{GPa}$	Noebe et al. (1993)
Burgers vector length of Ni-50Al	b	$\approx 5 \cdot 10^{-10}\,\mathrm{m}$	Noebe et al. (1993)
Constant	$\bar{\alpha}$	unity	Mecking and Kocks (1981)
Defect energy parameter	$k_\nabla = k_0\bar{\alpha}\mu b$	$27\,°\mathrm{C}$: $\approx 35 \cdot 10^{-6}\,\mathrm{N}$ with $k_0 = 1\,\mu\mathrm{m}$	-
		Interface	
Interface resistance parameter	k_Γ	$\approx 0.208\,\mathrm{Nm}^{-1}$	This study
	k_σ	$\approx 0.15\,\mu\mathrm{m}$	This study

8.3 Validation of the IGFEM approach in the context of gradient plasticity

The enriched shape functions described in Section 5.2.3 are able to discretize field variables containing weak discontinuities at interfaces. They were derived following the interface-enriched generalized finite element method (IGFEM) introduced by Soghrati et al. (2012). However, in order to validate this approach in the context of gradient plasticity, the simulation results using the enriched shape functions are compared to an analytical solution. Therefore, a periodic laminate under simple shear is considered, as depicted in Fig. 8.1 b). To obtain the exact analytical solution, stationary single slip ($\gamma_{\mathrm{acc}}(x) = \gamma(x)$) as well as homogeneous shear stress are assumed. Further, the ideal case of perfect coincidence between the micromorphic variable $\check{\gamma}(x)$ and its associated variable $\gamma(x)$ is considered, thus, $\check{\gamma}(x) = \gamma(x)$. The analytical solution is solved separately for each lamella β in its local coordinate system x_β, depicted in Fig. 8.1 b). To keep the analytical solution simple, the same material and the same lamella thickness is chosen for both lamellas. Consequently, no distinction between the two lamellas, if not necessary, will be made. In addition, the shear stress is assumed to be monotonic and positive ($\tau \geq 0$). The necessary equilibrium conditions as well as the jump conditions are summarized in Box 8.1 for the general three-dimensional case and simplified for the one-dimensional single-slip case to

$$\frac{\mathrm{d}\tau}{\mathrm{d}x} = 0, \quad \pi(x) - \frac{\mathrm{d}\xi(\mathrm{x})}{\mathrm{d}x} = 0, \quad \Xi = [\![\xi(x)]\!] n_\Gamma = (\xi^+ - \xi^-) n_\Gamma. \tag{8.24}$$

Here, n_Γ points from $\mathcal{V}^-$ to $\mathcal{V}^+$. The constitutive equations

$$\xi(x) = \frac{\partial W_\nabla}{\partial \mathrm{grad}\,(\check{\gamma}(x))} = k_\nabla \mathrm{grad}\,(\check{\gamma}(x))\,, \tag{8.25}$$

$$\pi(x) = \frac{\partial W_\chi}{\partial \check{\gamma}(x)} + \frac{\partial W_\mathrm{h}}{\partial \check{\gamma}(x)} = k_\mathrm{h}\check{\gamma}(x) + \tau_0 - \check{p}, \tag{8.26}$$

$$\Xi = k_\Gamma \tag{8.27}$$

are derived following the thermodynamically consistent approach by considering the dissipation inequality (described in Section 8.2.4 for the three-dimensional case) with the stored energy densities summarized in Box 8.2. Further, the reduced dissipation inequality motivates the flow rules for the bulk and for the interface as listed in Box 8.2. Here, contrary to the high temperature case, an isotropic hardening energy W_h was assumed. In the stationary single slip case (i.e. $\dot{\check{\gamma}}(x) = \dot{\gamma}(x) = 0$), the flow rule

$$\dot{\gamma}(x) = \dot{\gamma}_0 \left\langle \frac{\tau - \check{p}}{\tau^\mathrm{D}} \right\rangle^m \tag{8.28}$$

Box 8.2: Stored energy densities and flow rules for the numerical and analytical solution of a laminate under single slip to validate the IGFEM approach in the context of gradient plasticity.

Stored energy density	Flow rule
$W_\mathrm{e} = \frac{1}{2}\boldsymbol{\varepsilon}_\mathrm{e} \cdot \mathbb{C}[\boldsymbol{\varepsilon}_\mathrm{e}]$	$\dot{\gamma} = \dot{\gamma}_0 \left\langle \frac{\tau-\check{p}}{\tau^\mathrm{D}} \right\rangle^m$
$W_\nabla = \frac{1}{2}k_\nabla\,(\mathrm{grad}\,(\check{\gamma}) \cdot \mathrm{grad}\,(\check{\gamma}))$	$f_\Gamma := \Xi - k_\Gamma$ and
$W_\Gamma = k_\Gamma\check{\gamma}$ for $\boldsymbol{x} \in \Gamma$	$f_\Gamma \leq 0, \dot{\check{\gamma}} \geq 0, \dot{\check{\gamma}} f_\Gamma = 0$ for $\boldsymbol{x} \in \Gamma$
$W_\mathrm{h} = k_\mathrm{h}\,(\check{\gamma})^2 + \tau_0\check{\gamma}$	
$W_\chi = \frac{1}{2}H_\chi\,(\gamma_\mathrm{acc} - \check{\gamma})^2$	

leads to $\tau - \check{p} = 0$ and, thus, by using Eq. (8.25), Eq. (8.26), and the second equation in Eq. (8.24) to

$$\frac{\mathrm{d}^2}{\mathrm{d}x^2}\left(k_\nabla \check{\gamma}(x)\right) - k_\mathrm{h}\check{\gamma}(x) = \tau_0 - \tau. \tag{8.29}$$

By considering the two conditions

$$\left.\frac{\partial \check{\gamma}(x)}{\partial x}\right|_{x=0} = 0, \quad \check{\gamma}(x)|_{x=x_\Gamma} = \check{\gamma}_\Gamma, \tag{8.30}$$

the second-order differential equation can be solved to

$$\check{\gamma}(x) = \frac{g(x)}{g(x_\Gamma)}\left(\check{\gamma}_\Gamma - \frac{\tau - \tau_0}{k_\mathrm{h}}\right) + \frac{\tau - \tau_0}{k_\mathrm{h}}. \tag{8.31}$$

Here, x_Γ and $\check{\gamma}_\Gamma$ correspond to the interface's coordinate and the plastic slip of the interface, respectively. Further, for sake of improved readability, the abbreviation

$$g(x) = \exp\left(\sqrt{\frac{k_\mathrm{h}}{k_\nabla}}x\right) + \exp\left(-\sqrt{\frac{k_\mathrm{h}}{k_\nabla}}x\right) \tag{8.32}$$

was introduced. By using the two relationships

$$\left.\frac{\partial g(x_1)}{\partial x_1}\right|_{x_1=x_\Gamma} = -\left.\frac{\partial g(x_2)}{\partial x_2}\right|_{x_2=-x_\Gamma} := g'(x_\Gamma) \tag{8.33}$$

and $g(x_1 = x_\Gamma) = g(x_2 = -x_\Gamma)$, the jump condition can be reformulated to

$$\frac{-2k_\nabla g'(x_\Gamma)}{g(x_\Gamma)}\left(\check{\gamma}_\Gamma - \frac{\tau - \tau_0}{k_\mathrm{h}}\right) = k_\Gamma. \tag{8.34}$$

Finally, the plastic slip of each lamella β can be calculated by using Eq. (8.31) and the interface's plastic slip evolution

$$\check{\gamma}_\Gamma = \begin{cases} 0, & \text{for } \tau \leq \dfrac{k_\Gamma k_\mathrm{h} g(x_\Gamma)}{2k_\nabla g'(x_\Gamma)} + \tau_0 \\ \dfrac{\tau - \tau_0}{k_\mathrm{h}} - \dfrac{k_\Gamma g(x_\Gamma)}{2k_\nabla g'(x_\Gamma)}, & \text{otherwise.} \end{cases} \tag{8.35}$$

This analytical solution of a lamella with stationary single slip and homogeneous shear stress was already used by Bayerschen and Böhlke (2016) for microscopically hard interfaces.
All necessary assumptions to derive the analytical solution could be transferred to the simulation directly, except for the stationary condition. Consequently, to account for an approximation of the stationary condition, a low shear loading rate of $6 \cdot 10^{-6}\,\mathrm{MPa/s}$ was chosen (thus $300\,\mathrm{MPa}$ in $5 \cdot 10^{7}\,\mathrm{s}$). Isotropic elastic behavior is assumed for the lamella with the Poisson's ratio ν and the shear modulus μ. All material properties are summarized in Tab. 8.2 and were chosen in order to keep the error of the non-stationary slip small. The impact of the specific material properties on the error is given in Appendix A.3 together with an estimation of the difference between the stationary and non-stationary slip. The results of the simulation are depicted in Fig. 8.1 a) together with the analytical solution for specific shear stresses. The location of the interface boundaries are marked by the vertical dotted lines (at $\tilde{x} = \pm 0.25\,\mu\mathrm{m}$). For the stresses $\tau = 225\,\mathrm{MPa}$ and $\tau = 250\,\mathrm{MPa}$, the resistance of the interface is still sufficiently high to prevent plastic slip.

Table 8.2: Used parameters for the numerical and the analytical solution of a laminate.

Parameter	Value	Parameter	Value	Parameter	Value
k_∇	$35 \cdot 10^{-6}\,\mathrm{N}$	k_Γ	$10\,\mathrm{Nm}^{-1}$	k_h	$5 \cdot 10^{3}\,\mathrm{MPa}$
τ_0	$200\,\mathrm{MPa}$	ν	0.22	μ	$42\,\mathrm{GPa}$
τ^D	$50\,\mathrm{MPa}$	m	2	$\dot{\gamma}_0$	$5 \cdot 10^{-3}\,\mathrm{s}^{-1}$
H_χ	$10^{15}\,\mathrm{MPa}$				

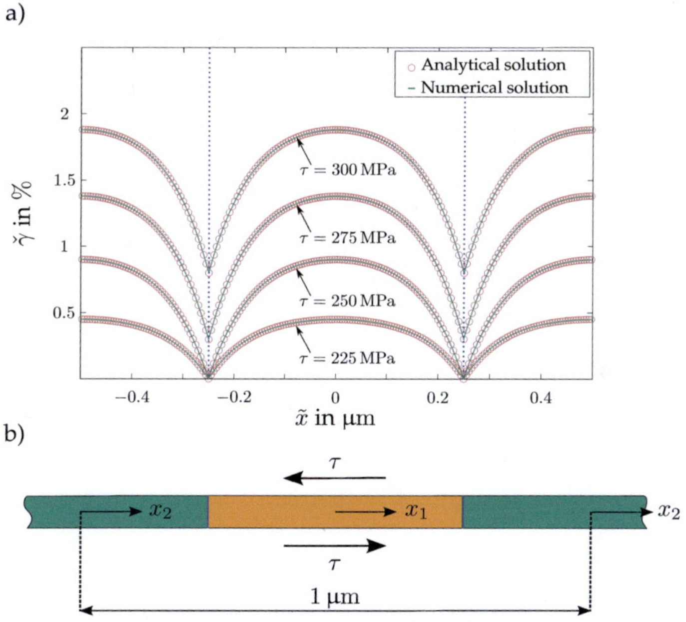

Figure 8.1: a) Comparison of the analytical results with the numerical ones of a periodic laminate for single slip. The plastic slip $\check{\gamma}$ of the numerical calculation is in perfect agreement with the analytical solution for various applied shear stresses τ. b) One-dimensional periodic laminate microstructure applied to a homogeneous shear stress τ with periodic boundary conditions. (Albiez et al. (2018) Fig. 3)

The increase from $\tau = 225\,\mathrm{MPa}$ to $\tau = 250\,\mathrm{MPa}$, however, increases the micromorphic stress $\boldsymbol{\xi}$, which scales with the jump of $\mathrm{grad}\,(\check{\gamma})$. A further increase of the shear stress induces a plastic flow of the interface, shown for $\tau = 275\,\mathrm{MPa}$ and $\tau = 300\,\mathrm{MPa}$. At these stress values, the interface flow rule is fulfilled. In addition, a constant interface resistance was chosen for this exemplary problem, leading to a constant jump of $\mathrm{grad}\,(\check{\gamma})$ across the interface (compare $\mathrm{grad}\,(\check{\gamma})$ at $\tau = 275\,\mathrm{MPa}$ and $\tau = 300\,\mathrm{MPa}$). In conclusion, both results show perfect agreement, therefore, the IGFEM approach with the enriched shape functions described in Section 5.2.3 and the numerical integration of Section 5.2.4 is applicable in the context of gradient plasticity.

8.4 Simulation results for creep compared to experimental data

8.4.1 Method for revealing the effects of the interface resistance and the slip transfer/activation

The interplay during deformation between the fiber and the matrix is schematically depicted in Fig. 8.2 for specific deformation states. Here, experimentally and numerically obtained observations from literature were combined. The fibers possess essentially no defects, consequently, the mechanical behavior is purely elastic in the beginning. The free plastic flow of the matrix, however, leads to a load transfer from the matrix to the fibers (indicated by the red curved arrows in Fig. 8.2).

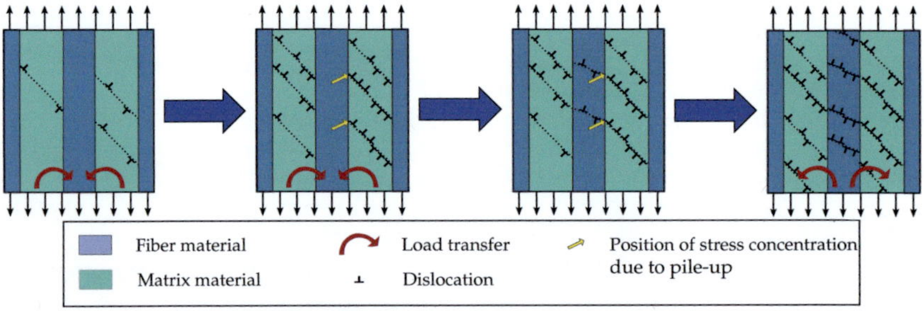

Figure 8.2: Schematical illustration of the suggested interplay between the fiber and the matrix during creep. From left to right: The free plastic flow of the matrix leads to a load transfer from the matrix to the fiber; the dislocations in the matrix accumulate at the fiber-matrix interface and form dislocation pile-ups; stress concentration due to the load transfer and the pile-up stress leads to the transfer/activation of slip in the fiber; reverse load transfer from the fiber to the matrix results from the softening of the fiber. (Albiez et al. (2018) Fig. 4)

This load transfer is macroscopically visible by a sharp reduction of the creep rate in the primary creep regime (Albiez et al., 2016a). Further, the GNDs accumulate in the matrix at the fiber-matrix interface due to the inhomogeneous deformation, which was experimentally observed as work hardening zones (WHZ) by Dudová et al. (2011). Shortly before reaching the minimum creep rate, the increased stress in the fiber due to the load transfer and the pile-up stress at the interface result in the transfer/activation of slip in the fiber. This was experimentally observed by scanning transmission electron microscopy investigations of Kwon et al. (2015) and is indicated with the yellow straight arrows in Fig. 8.2. The transfer/activation of slip in the fiber leads to the transition of an elastic to an elasto-plastic fiber's behavior and, therefore, induces a softening of the fiber. This is macroscopically observable as the increase of creep rate, thus, the passage through the minimum point in a creep curve. Furthermore, the dislocation density in the fiber increases quickly due to the creation and activation of further dislocation sources. As a result of the fiber's softening, a reverse load transfer from the fiber to the matrix takes place (marked by the red curved arrows in the sketch on the right in Fig. 8.2). This induces a further increase of the creep rate in the tertiary creep regime. In order to include these observations within the simulation model, the fiber is modeled purely elastic in the beginning. The matrix phase is described by an elasto-plastic behavior and, thus, induces a pile-up stress at the fiber-matrix interface due to the inhomogeneous deformation. However, as long as the pile-up stress does not reach the critical interface resistance value, no transfer/activation of slip in the fiber is permitted. Nevertheless, with increasing plastic deformation, the pile-up stress at the interface rises. By reaching the critical interface resistance and satisfying the loading condition, the interface flow rule is fulfilled leading to the transfer/activation of slip in the fiber. Consequently, this induces the transition of an elastic to an elasto-plastic fiber's behavior. This transition is induced abruptly and all over the fiber material. Furthermore, the dislocation multiplication modeled by Eq. (8.23) requires an initial dislocation density value ρ_0, therefore, the dislocation density is set to an almost negligible value (listed in Tab. 8.1) imitating an essentially defect-free state.

8.4.2 Boundary conditions and geometrical discretization of the microstructure

As stated in Section 7.2.5 the DS NiAl-9Mo microstructure allows the use of a representative unit cell. In the local simulation model of Albiez et al. (2016a), a unit cell with a hexagonal fiber arrangement was chosen. Compared to a quadratic fiber arrangement, the hexagonal arrangement leads to increased numerical costs, therefore, a quadratic one is chosen in this study and depicted in Fig. 8.3 a) (Albiez et al., 2018). The unit cell was discretized with an adaptive mesh with an increasing element-density towards the interface, shown in Fig. 8.3 b). The fiber spacing of $\lambda \approx 2\,\mu\mathrm{m}$ and the volume fraction of the fiber material (14 vol.-%) are kept consistent to the experimentally measured values. Hence, due to the link between the fiber spacing, the fiber diameter, and the volume fraction described by Eq. (2.3), only the fiber diameter is influenced by the microstructural change within the discretization. In addition, as shown in Fig. 8.3, the cross-section of the fiber is described as a perfect circle. To profit from the periodic microstructure, periodic boundary conditions for the displacement $\boldsymbol{u}$ and the plastic slip $\check{\gamma}$ were used within the finite element simulation. These boundary conditions were discussed in Section 5.3. The $\langle 100 \rangle$-orientation of both phases is aligned parallel to the growth direction (Bei and George, 2005; Dudová et al., 2011; Hu et al., 2013; 2012) and constant true stress $\bar{\sigma}$ was applied during creep simulation. The creep curves were obtained by calculating the Hencky strain based on the displacement of a master node. As discussed in Section 7.2.5, the determination of the exact zero point of plastic creep strain is difficult. Therefore, the simulated creep strain at the minimum point is set equal to the measured one.

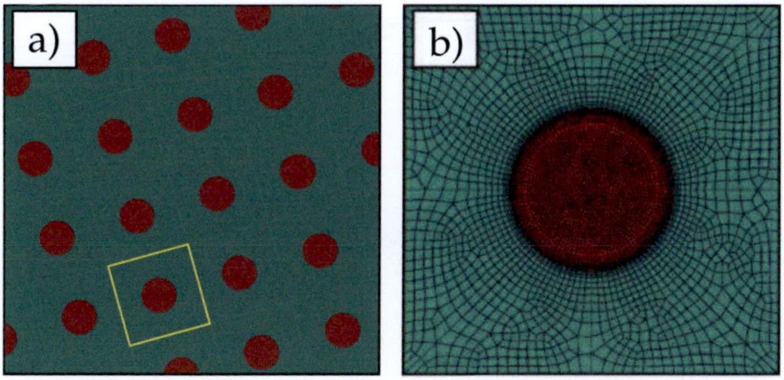

Figure 8.3: a) Geometrical discretization of the microstructure by assuming a perfect quadratic fiber arrangement. The representative unit cell is marked with the square. b) Representation of the mesh for the discretization of the unit cell, which includes approximately $2.7 \cdot 10^4$ nodes. (Albiez et al. (2018) Fig. 5 b) and c))

8.4.3 Material parameters for the matrix and the fibers of the DS NiAl-9Mo

The necessary material parameters for the simulation model of the DS NiAl-9Mo composite are physically motivated, thus, they are measurable or restricted to a reasonable range of physically motivated values. Therefore, experimental and numerical investigations out of literature provide most of the parameters (listed in Tab. 8.1). The elastic properties were calculated by the rule of mixture for the fiber and for the matrix, whereas the values are close to the experimental measured values of pure Mo and stoichiometric NiAl from Wern (2004) and Rusović and Warlimont (1977), respectively. Further, the impact of the elastic properties on the overall creep behavior is rather small (discussed in Section 7.4.6 and illustrated in Fig. 7.7). The reference shear rate of the fiber $\dot{\gamma}_0$, however, is not given in literature for a model considering inhomogeneous plastic deformation. Further, the two parameters for the interface resistance k_Γ as well as k_σ are not given.

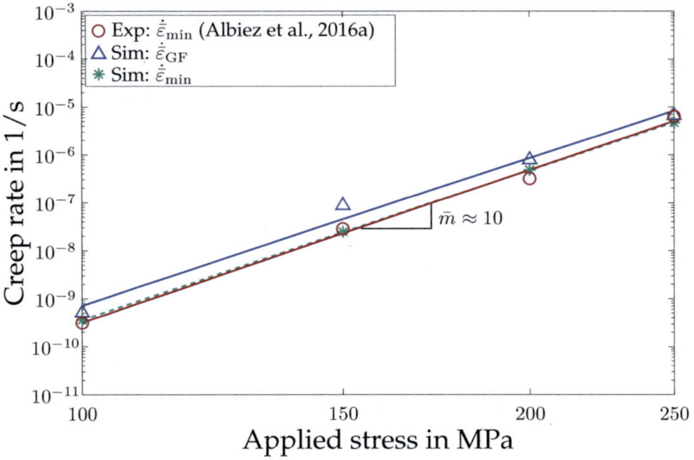

Figure 8.4: Simulated and experimentally measured minimum creep rates of the composite ($\dot{\bar{\varepsilon}}_{\min}$) vs. the applied stress. The creep rates $\dot{\bar{\varepsilon}}_{\mathrm{GF}}$ where the interface flow rule has been fulfilled for the first time are marked by the triangles. The interface flow rule is fulfilled shortly before reaching the minimum creep rate, initiating softening of the fiber. The experimental results at an applied stresses of 100 MPa and 250 MPa were used to identify the missing material parameters, however, the results for the applied stresses of 150 MPa and 200 MPa were predicted by the simulation model. Both, the stress exponent of the simulation and the experimentally measured one are in perfect agreement. (Albiez et al. (2018) Fig. 6)

In order to identify the missing material parameters, the simulation was fitted to the creep experiments from Albiez et al. (2015; 2016a) at 1000 °C for an applied stress of 250 MPa and 100 MPa, respectively. The details of the experiments and the sample preparation are given in Albiez et al. (2016a). The simulated and the experimental minimum creep rate $\dot{\bar{\varepsilon}}_{\min}$ are in good agreement for both applied loads (shown in Fig. 8.4). Further, the experimentally measured creep curve and the simulated one coincide for an applied stress of 250 MPa (depicted in Fig. 8.5).

8.4.4 Evaluation of the interface flow rule

The stress dependent interface flow rule (Eq. (8.19)) could be motivated by several investigations, e.g., De Koning et al. (2002); Dewald and Curtin (2007); Tschopp et al. (2008); Bachurin et al. (2010) and was discussed in Section 8.2.5. It can be evaluated by comparing the experimentally measured minimum creep rates with the predicted ones for specific applied stresses. Shortly before reaching the minimum creep rate $\dot{\varepsilon}_{\mathrm{min}}$, the interface flow rule should be fulfilled, initiating the transfer/activation of slip in the fiber, which can be interpreted as transfer/nucleation of dislocations. As a consequence of those dislocations, the fiber material softens and, thus, the creep rate of the composite increases. This was emphasized in Section 8.4.1 and illustrated by the scheme in Fig. 8.2. In the work at hand, however, no balance equation for the dislocations is used (contrary to the work of Wulfinghoff and Böhlke (2015)). Consequently, a distinction between transfer and activation of dislocations at the interface is not possible.

The material parameters for the interface resistance (k_Γ and k_σ) were identified by fitting the simulations for applied stresses of $100\,\mathrm{MPa}$ and $250\,\mathrm{MPa}$ to the corresponding creep experiments. For intermediate applied stresses the minimum creep rates were predicted and are in perfect agreement with the experimentally observed ones (shown in Fig. 8.4). Furthermore and most importantly, the interface flow rule is fulfilled shortly before reaching the minimum creep rate for all investigated applied stresses. This is shown in Fig. 8.4 by the creep rate of the composite in the moment of fulfilling the interface flow rule for the first time (denoted by $\dot{\varepsilon}_{\mathrm{GF}}$). Consequently, one can conclude that the gradient plasticity model and the stress dependent interface flow rule reproduce well the experimental observations.

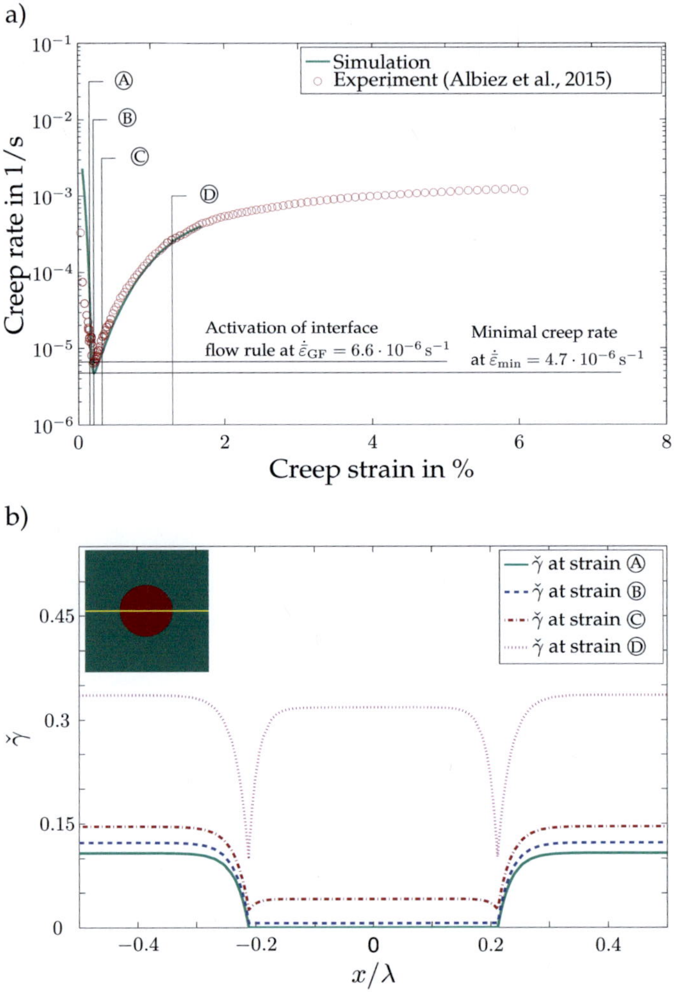

Figure 8.5: a) Simulated and experimentally measured creep curve at 1000 °C, an applied stress of 250 MPa, and a fiber diameter of $d = 0.729\,\mu\mathrm{m}$ (Albiez et al., 2018). The comparison shows a perfect agreement. The interface flow rule is fulfilled shortly before reaching the minimum creep rate, leading to the softening of the fiber. b) Micromorphic plastic slip $\check{\gamma}$ curves for a cut through the microstructure at different macroscopic strain values which are depicted in a). The position of the cut is highlighted in the left top corner by the continuous yellow line. (Albiez et al. (2018) Fig. 7)

8.5 Discussion of the gradient crystal plasticity model

The above performed evaluation clearly shows the capability of the gradient crystal plasticity model with the stress dependent interface flow rule to reproduce the mechanical behavior of the DS NiAl-9Mo composite. Within this section, the gained insights into the composite's behavior due to the simulation model will be discussed and summarized. The schematically depicted interplay between the fibers and the matrix, shown in Fig. 8.2, is well reproduced by the model. This suggested interplay arises due to experimental and numerical investigations from the literature, however, it was not fully validated up to now. This interplay can be further reviewed by analyzing the micromorphic plastic slip $\check{\gamma}$ distribution within the composite. Therefore, Fig. 8.5 b) shows the micromorphic plastic slip $\check{\gamma}$ for a cut through the microstructure, whereas the position of the cut is depicted by the yellow continuous line in the left top corner of Fig. 8.5 b). Each of the $\check{\gamma}$ curves in Fig. 8.5 b) can be related to a macroscopic strain value of the composite Ⓐ-Ⓓ, which are marked in the creep curve in Fig. 8.5 a) for sake of improved comprehension. In the primary creep regime (at strain Ⓐ), the interface flow rule has not been fulfilled yet. Consequently, due to the defect-free as-grown state of the fiber, they behave still purely elastic ($\check{\gamma}$=0). Contrary, the matrix flows freely (experimentally observed by Bei et al. (2008a;b; 2005)) but inhomogeneously. As an indicator for the GNDs, the gradient of the micromorphic plastic slip ($\operatorname{grad}(\check{\gamma})$) can be used, which has its maximum at the fiber-matrix interface. The accumulation of GNDs at the interface can be interpreted as pile-up, inducing stresses at the interface. If these pile-up stresses reach the interface resistance and the loading condition is fulfilled, the dislocation transfer/nucleation in the fiber is activated. This was experimentally observed by Kwon et al. (2015) in form of Hairpin-like dislocations nucleating at stress concentrations at the fiber-matrix interface. Strain Ⓑ corresponds to the point of minimum creep rate. Here, the interface flow rule has already been fulfilled for the first time. Consequently, a small amount of defects are present in the fiber, thus, the behavior of the fiber changed from purely elastic to elasto-plastic. Those defects soften the fiber (described by Eq. (8.22)) and from now on the creep rate starts to increase. The softening behavior results from the transition from a dislocation-source strengthening mechanism to a forest-dominated strengthening mechanism with increasing dislocation density (discussed in Chapter 6). Even though the micromorphic plastic slip $\check{\gamma}$ in the fiber is still small and quite homogeneous at strain Ⓑ, it increases and starts to generate GNDs at the interface. This is visible for the $\check{\gamma}$ curve at strain Ⓒ in Fig. 8.5 b). The accumulation of GNDs on both sides of the interface due to its resistance is shown for the final strain Ⓓ and was experimentally observed by Barabash et al. (2010a). To analyze the load transfer between both phases, the calculated Cauchy stress $\boldsymbol{\sigma}$ can be decomposed into its eigenvalues $\hat{\sigma}_i$ and eigenvectors $\hat{\mathbf{e}}_i$, thus,

$$\boldsymbol{\sigma} = \sum_{i=1}^{3} \hat{\sigma}_i \hat{\mathbf{e}}_i \otimes \hat{\mathbf{e}}_i. \tag{8.36}$$

It is possible to find one eigenvector lying parallel to the fiber's direction at each point within the simulation, whereas the other two eigenvectors are positioned in the plane perpendicular to the fiber's direction. The eigenvetor parallel to the fiber's direction and the corresponding eigenvalue are denoted by $\hat{\mathbf{e}}^{||}$ and $\hat{\sigma}^{||}$, respectively. Within the simulations and the experiments the load was applied parallel to the fiber's direction, consequently, the stress $\hat{\sigma}^{||}(\hat{\mathbf{e}}^{||} \otimes \hat{\mathbf{e}}^{||})$ carries most of the applied stress. The absolute value of this stress is shown in Fig. 8.6 a) for the different strains Ⓐ-Ⓓ, where due to symmetry only a quater of the microstructure is depicted. An inhomogeneous stress distribution in the matrix with the maximal value at the fiber-matrix interface is visible in the beginning (at strain Ⓐ depicted in the left top corner of Fig. 8.6 a)). It results from the inhomogeneous plastic deformation in the matrix with the accumulation of GNDs at the interface. The fiber, however, is essentially defect free and, thus, behaves purely elastic. Consequently, the stress is homogeneously distributed and is obviously greater than the stress in the matrix. The subfigure depicted in the right top corner of Fig. 8.6 a) shows $|\hat{\sigma}^{||}|$ at the minimum creep rate (strain Ⓑ). Here, the interface flow rule was fulfilled, inducing defects in the fiber and changing the fiber's behavior from purely elastic to elasto-plastic. The stress distribution of the fiber contains already small inhomogeneities, however, they are more apparent at strain Ⓒ. Finally, at strain Ⓓ, the inhomogeneous stress distribution is visible for both phases with increasing values towards the interface.

To explain the shape of the creep curve, the ratio of the load carried by the fiber $\bar{\sigma}_{\mathrm{f}}^{||}$ to the load carried by the matrix $\bar{\sigma}_{\mathrm{m}}^{||}$ is shown in Fig. 8.6 b), with

$$\bar{\sigma}_{\mathrm{f}}^{||} = \frac{1}{V_{\mathrm{c}}} \int_{V_{\mathrm{f}}} \hat{\sigma}^{||} \mathrm{d}v, \qquad \bar{\sigma}_{\mathrm{m}}^{||} = \frac{1}{V_{\mathrm{c}}} \int_{V_{\mathrm{m}}} \hat{\sigma}^{||} \mathrm{d}v, \tag{8.37}$$

so that $\bar{\sigma} = \bar{\sigma}_{\mathrm{f}}^{||} + \bar{\sigma}_{\mathrm{m}}^{||}$. Here, the composite's volume $V_{\mathrm{c}} = V_{\mathrm{f}} + V_{\mathrm{m}}$ is composed of the fiber's volume V_{f} and the matrix's volume V_{m}. Until strain Ⓑ is reached, the ratio $\bar{\sigma}_{\mathrm{f}}^{||}/\bar{\sigma}_{\mathrm{m}}^{||}$ increases indicating the load transfer from the matrix to the fiber (Fig. 8.6 b)). This load transfer leads to a decrease of the creep rate to the minimum point (depicted in Fig. 8.5 a)). At the minimum creep rate (strain Ⓑ) the maximal value of $\bar{\sigma}_{\mathrm{f}}^{||}/\bar{\sigma}_{\mathrm{m}}^{||}$ is reached. Here, the load carried by the fiber is approx. 6.5 times the load carried by the matrix, which agrees with the observations of Bei et al. (2005). They stated that the fiber phase essentially carries the entire applied load. By considering the volume fraction of the fiber, it can be shown that the mean stress in the fiber is approx. 6 times higher than the applied stress. This is in good agreement with the results of the shear lag model of Hu et al. (2013), which revealed that the mean stress in the fiber is approx. 7 times higher than the applied stress. The interface flow rule was fulfilled for the first time shortly before reaching strain Ⓑ, leading to the softening of the fiber. As a result of the loss of fiber's strength, a reverse load transfer from the fiber to the matrix takes place, shown in Fig. 8.6 b) by the decrease of the $\bar{\sigma}_{\mathrm{f}}^{||}/\bar{\sigma}_{\mathrm{m}}^{||}$-ratio. The composite's creep resistance decreases by the fiber's softening, leading to an increase of the composite's creep rate in the tertiary creep regime. In summary, one can conclude that the composite's creep behavior strongly depends on the load partitioning between both phases. This is in line with the observations from Barabash et al. (2010b). Further, the analysis of the load partitioning between both phases helps to explain the shape of the creep curve. Finally, the simulated creep agrees well with the experimentally measured one (see Fig. 8.5 a)), leading to the conclusion that the model represents the most important characteristics of the composite quite well.

The stress dependent interface flow rule (Eq. (8.19)) was motivated in Section 8.2.5 by considering the dissipation inequality of the interface. The two necessary parameters (k_Γ and k_σ) were identified by fitting the simulated minimum creep rate to the experimental one for applied stresses of $100\,\mathrm{MPa}$ and $250\,\mathrm{MPa}$. In Fig. 8.4, however, the predicted results for applied stresses of $150\,\mathrm{MPa}$ and $200\,\mathrm{MPa}$ are compared with experimental results. Also for the predicted creep curves, the interface flow rule is fulfilled for the first time shortly before reaching the minimum creep rate, as shown in Fig. 8.4. Here, the triangles correspond to the creep rates of the composite $\dot{\bar{\varepsilon}}_{\mathrm{GF}}$ where the interface flow rule has been fulfilled for the first time. Those creep rates are close to the simulated and measured minimum creep rates, depicted by a star and a circle, respectively. Consequently, for all investigated applied stresses, one can conclude that shortly before reaching the minimum creep rate the interface flow rule is fulfilled and initiates the softening of the fiber.

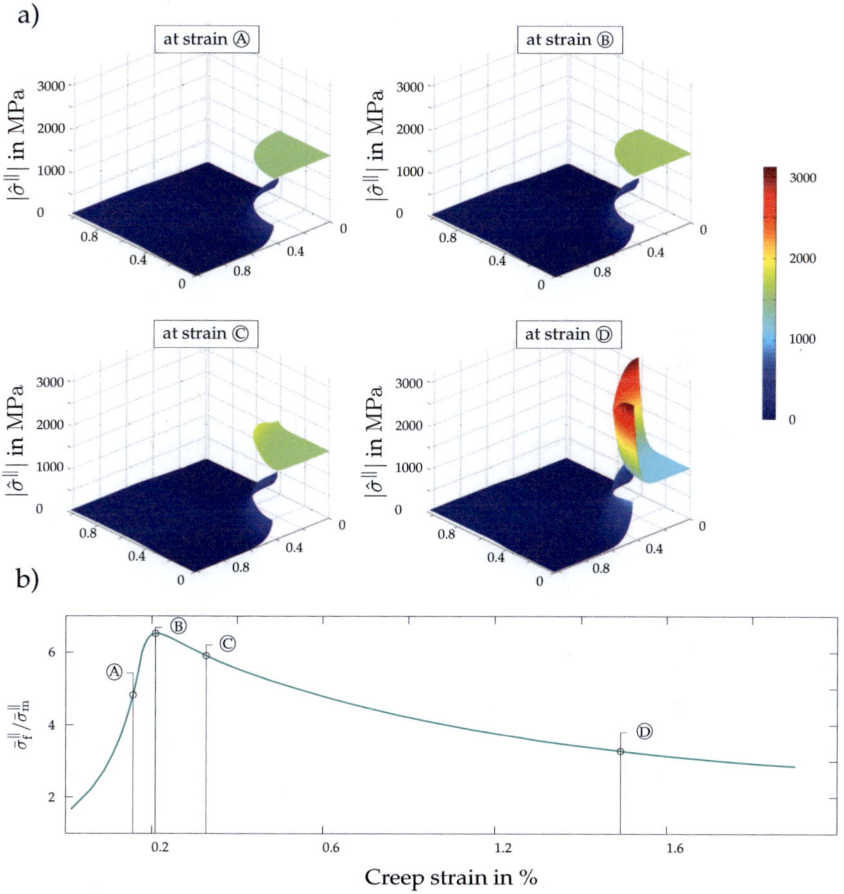

Figure 8.6: a) $\hat{\sigma}^{||}$ at different macroscopic strain values which are depicted in Fig. 8.5 a). Due to symmetry, only a quater of the microstructure is shown for sake of improved illustration. The fiber supports higher loads compared to the matrix for all considered strain values. The initial homogeneous distribution of $\hat{\sigma}_f^{||}$ in the fiber results from the linear-elastic behavior and changes to an inhomogeneous one at higher strains. b) The ratio $\bar{\sigma}_f^{||}/\bar{\sigma}_m^{||}$ illustrates the load partitioning between the fiber and the matrix during the creep deformation. (Albiez et al. (2018) Fig. 8)

The calculated stress exponent of the composite $\bar{m} \approx 10.4$ corresponds to the slope of the linear fit of $\dot{\bar{\varepsilon}}_{\min}$ and is close to the slope $\bar{m}_{\mathrm{GF}} \approx 10.2$ of $\dot{\bar{\varepsilon}}_{\mathrm{GF}}$. Both stress exponents are in the range of the experimentally measured values of pure Mo ($m_{\mathrm{Mo}} \approx 10.5$). Consequently, in addition to the statement of Albiez et al. (2016a) that the creep behavior of the composite is mainly controlled by the fibers, one may conclude that the fiber's interface behavior is important, too. This is in agreement with the work of Barabash et al. (2010b), who stated that the fiber-matrix interface plays a particularly crucial role. Furthermore, it is noteworthy that by neglecting the stresses at the interface and, thus, using a constant interface flow rule, the model would not lead to $\bar{m} \approx \bar{m}_{\mathrm{GF}}$. Therefore, considering the stresses at the interface is necessary to describe the composite's behavior.

The mutual dislocation annihilation term $k_2 \rho_{\mathrm{ssd}}$ within the dislocation evolution equation (8.23) is questionable (Albiez et al., 2016a) and, thus, was not considered within the simulations. Analyzing the ratio of the dislocation evolution rate with and without the mutual dislocation annihilation, thus,

$$\begin{aligned} \frac{\dot{\rho}^{\mathrm{k1}}_{\mathrm{ssd}} - \dot{\rho}^{\mathrm{k12}}_{\mathrm{ssd}}}{\dot{\rho}^{\mathrm{k1}}_{\mathrm{ssd}}} &= \frac{(k_1 \sqrt{\rho_{\mathrm{ssd}}})\dot{\bar{\gamma}} - (k_1 \sqrt{\rho_{\mathrm{ssd}}} - k_2 \rho_{\mathrm{ssd}})\dot{\bar{\gamma}}}{(k_1 \sqrt{\rho_{\mathrm{ssd}}})\dot{\bar{\gamma}}} \\ &= \frac{k_2}{k_1} \sqrt{\rho_{\mathrm{ssd}}} \end{aligned} \tag{8.38}$$

can help to rationalize this simplification. Albiez et al. (2016a) identified the dislocation annihilation parameter as $k_2 = 66$. Together with the dislocation multiplication parameter k_1 from Tab. 8.1 the ratio is obtained to

$$\frac{\dot{\rho}^{\mathrm{k1}}_{\mathrm{ssd}} - \dot{\rho}^{\mathrm{k12}}_{\mathrm{ssd}}}{\dot{\rho}^{\mathrm{k1}}_{\mathrm{ssd}}} \approx 14\,\%. \tag{8.39}$$

Here, the calculated mean statistically stored dislocation density value of $\rho_{\mathrm{ssd}} = \bar{\rho}^{\mathrm{f}}_{\mathrm{ssd}}$ within the fiber at strain © was used. In the considered case, the deviation is rather small, thus, it was reasonable to neglect the dislocation annihilation term. This deviation, however, increases with increasing plastic strain. Consequently, the mutual dislocation annihilation term should be taken into account in the future. Especially simulations analyzing the ternary creep regime where higher creep strains are necessary, should incorporate this effect.

The consideration of a hexagonal arrangement of the fibers increases the numerical costs, therefore, a quadratic arrangement was chosen. Due to this microstructural change, an adjustment of the fiber's diameter, the fiber spacing, or the volume fraction of the fiber is necessary. The simulation results depicted in Fig. 7.11 reveal the impact of the fiber's volume fraction to the creep behavior, thus, the volume fraction was kept consistent with the experimentally observed value. In addition, the fiber spacing might also affect the minimum creep rate due to the inhomogeneous plastic deformation within the matrix phase (shown in Fig. 8.6 a) for strain Ⓑ). As a result, the fiber's diameter was reduced by a factor of $3^{-1/4} \approx 0.76$ in the discretization of the unit cell. It was shown that the minimum creep rate of the composite is unaffected by a reduction of the fiber's diameter by a factor of approx. 0.64, as discussed in Section 7.4.4 and depicted in Fig. 7.5. This can be rationalized by the fact that until the interface flow rule is fulfilled, the fiber behaves purely elastic, thus, independent of its diameter. Consequently, the modification of the fiber's diameter does not strongly affect the simulation results in the considered range and a quadratic arrangement can be used to analyze the interface flow rule. The creep rate in the tertiary creep regime, however, depends on the fiber diameter. Therefore, to further investigate this regime, a hexagonal arrangement should be used for a validation of the model, especially in the case of different microstructures.

The defect energy parameter k_∇ can be estimated based on the self-energy of GNDs ($E_0 = \alpha \mu b^2$) (Ohno and Okumura, 2007; Hull and Bacon, 2011). However, this is only valid in the case of a defect energy density which is linear to the norm of the gradient plastic slip, thus, linear to $||\mathrm{grad}\,(\check{\gamma})\,||$. The linear dependency leads to a constant absolute value of the gradient stress and, thus, to a gradient stress unaffected by the amount of GNDs. Therefore, in this work, a quadratic defect energy is considered, leading to a linear dependency of the gradient stress to the geometrically necessary dislocation density. However, to enable the use of the self-energy of GNDs, a normalization constant k_0 was used as suggested in the work of Bayerschen and Böhlke (2016). For the evaluation of the defect energy parameter k_∇, the calculated size of the work hardening zones are compared to the experimentally observed one. A high/low defect energy parameter k_∇ leads to a wide/narrow work hardening zone and a low/high geometrically necessary dislocation density, respectively. The relative geometrically necessary dislocation densities are shown in Fig. 8.7 a) for strain Ⓓ and are given by

$$\rho_{\mathrm{rel}}^{\mathrm{f/m}} = \frac{\rho_{\mathrm{GND}}^{\mathrm{f/m}}}{\bar{\rho}_{\mathrm{GND}}} = \frac{1}{b^{\mathrm{f/m}}\bar{\rho}_{\mathrm{GND}}}\sqrt{|\mathrm{grad}\,(\check{\gamma})\cdot\mathrm{grad}\,(\check{\gamma})\,|}. \tag{8.40}$$

Here, $\bar{\rho}_{\mathrm{GND}}$ corresponds to the mean value of ρ_{GND} for the whole structure. Further, the relative geometrically necessary dislocation densities of the fiber and the matrix are denoted by $\rho_{\mathrm{rel}}^{\mathrm{f}}$ and $\rho_{\mathrm{rel}}^{\mathrm{m}}$, respectively. The relative geometrically necessary dislocation density of each phase increases towards the fiber-matrix interface Γ, where its maximum values are reached. The yellow dotted circles indicate the location where the density value ρ_{GND} corresponds to the mean value, thus, $\rho_{\mathrm{GND}} \approx \bar{\rho}_{\mathrm{GND}}$. At each point in between these two circles, the density value is higher than the mean value, making the enclosed area to an approximation of the work hardening zones. To evaluate the calculated GND-field and its work hardening zone, they are compared to the transmission electron microscopy (TEM) bright-field image of Dudová et al. (2011), shown in Fig. 8.7 b). Here, the dislocations of a sample subjected to $35\,\%$ compression creep strain at $900\,^{\circ}\mathrm{C}$ and an applied stress of $300\,\mathrm{MPa}$ are shown. Despite this different creep test conditions, the similarities of the calculated work hardening zone and the experimentally observed one are obvious. Furthermore, Hu et al. (2013) investigated the work hardening zone of a sample deformed at an applied stress of $150\,\mathrm{MPa}$ and at $1100\,^{\circ}\mathrm{C}$ near to the minimum creep strain ($0.3\,\%$) and measured a dislocation density of $\approx 2.5 \cdot 10^{15}\,\mathrm{m}^{-2}$. A local dislocation density of $1.77 \cdot 10^{14}\,\mathrm{m}^{-2}$ within the work hardening zone was observed by Dudová et al. (2011) (sample subjected to $33\,\%$ compression creep strain at $900\,^{\circ}\mathrm{C}$ and an applied stress of $200\,\mathrm{MPa}$). Both experimental dislocation measurements are in line with the calculated maximum value of GNDs in the matrix ($\max(\rho^{\mathrm{m}}_{\mathrm{GND}}) \approx 10^{15}\,\mathrm{m}^{-2}$) at strain Ⓓ, even though the sample deformation differs. Consequently, due to both agreements, the size of the work hardening zone and the absolute value of GNDs within the work hardening zone, one can conclude that the defect energy parameter k_{∇} is reasonable.

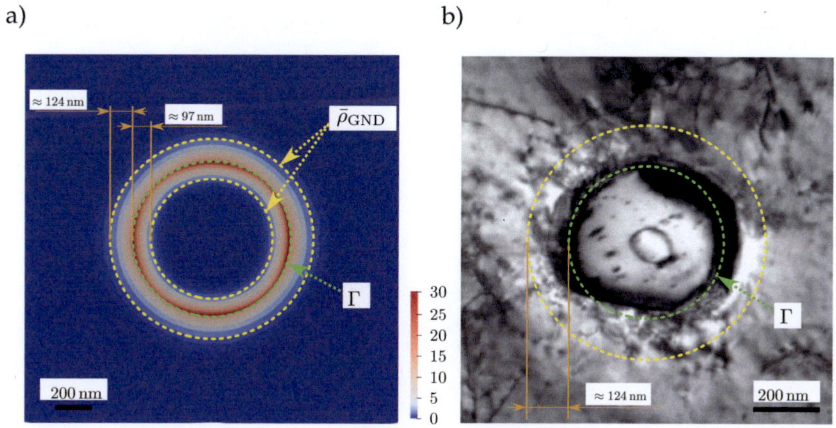

Figure 8.7: a) Calculated geometrically necessary dislocation density normalized by the mean value $\bar{\rho}_{\mathrm{GND}}$, indicating the dislocation accumulation around the fiber for a compression creep strain value of $< 2\,\%$. A higher density value than the mean value is observed within the two yellow dotted circles, thus, the GNDs accumulate near the interface Γ. b) Experimentally observed dislocation accumulation around the fiber after $35\,\%$ compression creep strain at $900\,^{\circ}\mathrm{C}$ (TEM bright-field image out of the work from Dudová et al. (2011)). (Albiez et al. (2018) Fig. 9)

8.6 Summary and conclusion of the gradient crystal plasticity model

The proposed gradient crystal plasticity model extends the local crystal plasticity model introduced in Chapter 7 by incorporating:

- The impact of geometrically necessary dislocations in terms of the gradient field ($\mathrm{grad}\,(\check{\gamma})$) due to the inhomogeneous plastic deformation.
- The stress dependency of the slip transfer/activation at the fiber-matrix interface.
- The initial elastic behavior of the fiber due to its essentially defect-free state.

Further, the flow rule for the bulk as well as the interface flow rule were derived in a thermodynamically consistent framework. As a result, the model incorporates more physical aspects compared to the local crystal plasticity model introduced in Chapter 7 and, thus, represents an improvement. The weak discontinuity of the $\check{\gamma}$ field at the fiber-matrix interface was considered by the enriched shape functions introduced in Section 8.3. This approach was validated in the context of gradient plasticity by comparing the simulation results of a perfect lamella with the corresponding analytical solution. The model for the DS NiAl-9Mo was evaluated by comparing the simulated results to experimentally obtained ones from literature. Due to the physical motivation of the simulation model, most of the model parameter were available in literature. It was shown that the simulated creep curve is in good agreement with the measured creep curve. In addition, the predicted minimum creep rates for different applied stresses are in line with the experimental findings, revealing the need of a stress dependent interface flow rule. Finally, by comparing the calculated size of the work hardening zones and their absolute dislocation density values to the investigations of Dudová et al. (2011); Hu et al. (2013), one can conclude that the simulation model which accounts for the stress dependent interface flow rule represents well the behavior and the characteristics of the composite. It was shown that in addition to the fiber phase, the fiber-matrix interface controls the composite's behavior. With the newly provided insights into the composite by the simulation model, the experimental findings from literature as well as the shape of the creep curve could be explained. For a closer analysis of the interfaces' behavior at elevated temperatures including the transition/activation of slip, additional simulations and experiments at various temperatures and with different microstructures are necessary. For this purpose, directionally solidified NiAl reinforced by a refractory metal, e.g., Cr, Mo is suitable, due to its regular microstructure and its adjustable fiber diameter/spacing.

Chapter 9

Coupling of balance equation of linear momentum with Cahn-Hilliard diffusion

9.1 Introduction and motivation

In the previous chapter, the interplay of dislocations with microstructural constraints, namely the fiber-matrix interface, was discussed. Hereby, the fiber diameter and the fiber spacing as the characteristic sizes of the microstructure were considered to be constant during the simulations and the experiments. This assumption is valid in the case of the directionally solidified NiAl-9Mo due to the eutectic composition and its thermodynamic stability (Johnson et al., 1995). However, several materials exhibit a change of the microstructure over time, especially at elevated temperatures. One of these materials is the Fe-61Al which contains a lamellar based microstructure (e.g., Scherf et al. (2015)). The orientation relationship between the two lamellar phases (FeAl and $FeAl_2$) was investigated in Bastin et al. (1978); Scherf et al. (2016). Li et al. (2017) calculated the lamellar coarsening of this composite by a phase field model and compared the results with experimental observations. The experimental findings suggest that the lamellar spacing λ increases with time t as

$$\lambda^3 = \lambda_0^3 + \dot{k}t, \tag{9.1}$$

with the rate constant $\dot{k}$ and the initial lamellar spacing λ_0. One may suggest that the lamellar coarsening influences the creep behavior, as observed in the case of γ-TiAl alloys (Appel et al., 2000; Wen et al., 2000). This was later investigated in Schmitt et al. (2017) and it was concluded that the instability of the microstructure influences the creep behavior.
To describe the coarsening process of the microstructure, the Cahn-Hilliard approach can be used. Contrary to the Fick's law, which leads to a homogeneous concentration distribution, the equilibrium state in the Cahn-Hilliard approach is reached for a homogeneously distributed chemical potential. This implies not necessarily a uniform concentration distribution (e.g., Ubachs et al. (2004)). The approach is based on a free energy which takes not only the local order parameter into account, but also the value of this parameter in its neighborhood through a gradient energy (Cahn and Hilliard, 1958). Gurtin (1996) introduced a Cahn-Hilliard equation based on a microforce balance. An extension of the classical Cahn-Hilliard model is proposed in Ubachs et al. (2004), where a strongly non-local variable was introduced which reduces the original fourth order partial differential equation into two second order ones. This enables the use of C_0-continous shape functions within the finite element context (e.g., Peerlings et al. (1996)). The approach was later used in the finite deformation framework for the describtion of the visco-plastic behavior of tin-lead solder (Ubachs et al., 2005). It is possible to derive the framework of Ubachs et al. (2004) by the micromorphic approach (Forest, 2009), too. A comparison between the micromorphic approach and the phase-field approach is given in Forest et al. (2011). Kaessmair and Steinmann (2016) analyzed higher-order methods to solve the Cahn-Hilliard equation, whereas Ammar et al. (2014) investigated the inheritance of plastic deformation within a phase field model.
In this chapter, the strongly non-local Cahn-Hilliard equation will be derived in a micromorphic framework. Further, the impact of the material parameters on the coarsening process and the equilibrium

state will be shortly investigated. The main part of the work focuses on the coupling between the diffusion equation and the balance of linear momentum through the elastically stored energy. Here, the impact of an inhomogeneous strain distribution on the coarsening process will be analyzed.

9.2 Equilibrium conditions and constitutive relationships

9.2.1 Diffusion equation

The diffusion equation can be derived by considering the mass balance per component α in a material volume

$$\int_{\mathcal{V}} \frac{\partial \varrho_\alpha}{\partial t} \mathrm{d}v + \int_{\mathcal{A}} \varrho_\alpha \boldsymbol{v}_\alpha \cdot \boldsymbol{n} \mathrm{d}a = 0. \tag{9.2}$$

Here, ϱ_α and $\boldsymbol{v}_\alpha$ correspond to the mass density and the velocity of the component α, respectively. By using the total mass density of the mixture with N_c components

$$\varrho = \sum_{\alpha=1}^{N_\mathrm{c}} \varrho_\alpha \tag{9.3}$$

and the barycentric velocity

$$\bar{\boldsymbol{v}} = \sum_{\alpha=1}^{N_\mathrm{c}} \frac{\varrho_\alpha \boldsymbol{v}_\alpha}{\varrho}, \tag{9.4}$$

the total mass balance is obtained by summation over all components, thus,

$$\int_{\mathcal{V}} \frac{\partial \varrho}{\partial t} \mathrm{d}v + \int_{\mathcal{A}} \varrho \bar{\boldsymbol{v}} \cdot \boldsymbol{n} \mathrm{d}a = 0. \tag{9.5}$$

Introducing the mass concentration $c_\alpha = \varrho_\alpha/\varrho$ and substituting Eq. (9.5) in Eq. (9.2) results in the diffusion equation of the component α

$$\int_{\mathcal{V}} \varrho \dot{c}_\alpha \mathrm{d}v = -\int_{\mathcal{A}} \varrho_\alpha (\boldsymbol{v}_\alpha - \bar{\boldsymbol{v}}) \cdot \boldsymbol{n} \mathrm{d}a = -\int_{\mathcal{A}} \boldsymbol{j}_\alpha \cdot \boldsymbol{n} \mathrm{d}a. \tag{9.6}$$

Here, the diffusion mass flow $\boldsymbol{j}_\alpha$ of the component α was introduced. By using the transformation relationships between the current and the reference placement, summarized in Box 4.1, the diffusion equation in the reference placement is given by

$$\int_{\mathcal{V}_0} \dot{c}_\alpha \mathrm{d}V = -\int_{\mathcal{V}_0} \mathrm{Div}\left(\tilde{\boldsymbol{j}}_\alpha\right) \mathrm{d}V \tag{9.7}$$

where the abbreviation $\tilde{\boldsymbol{j}}_\alpha = \varrho^{-1}\boldsymbol{F}^{-1}\boldsymbol{j}_\alpha$ is used.

9.2.2 Equilibrium equations for the strongly non-local Cahn-Hilliard diffusion

Following the approach of Forest (2009), the additional necessary equilibrium equation for the Cahn-Hilliard model is derived by a micromorphic approach. Therefore, the general procedure described in Section 4.5 is used. Here, a two-phase material is considered, resulting in an improved readability with $c_1 = c$ and $c_2 = 1 - c$, due to $c_1 + c_2 = 1$. In the Cahn-Hilliard model, the general internal variable ϕ corresponds to the mass concentration c and the general micromorphic variable $\check{\phi}$ to $\check{c}$ (also called microconcentration in, e.g., Forest (2018)). The micromorphic variable can be regarded as a description of the composition variance inside the volume element (Forest, 2018). Contrary to the previous chapter, the interface Γ is considered as a diffuse interface. All necessary conditions for the Cahn-Hilliard model are summarized in Box 9.1, whereas the suffix $(\cdot)_\mathrm{G}$ indicating the general variables, was replaced by $(\cdot)_\mathrm{d}$.

Box 9.1: Equilibrium conditions and boundary conditions for the coupled Cahn-Hilliard framework of a two-phase material.

Field variable	Equilibrium condition $\forall \boldsymbol{x} \in \mathcal{V}$	Boundary condition on $\mathcal{A}_\mathrm{t} \vee \mathcal{A}_\Xi$
$\boldsymbol{u}$	$\operatorname{div}(\boldsymbol{\sigma}) = \boldsymbol{0}$	$\boldsymbol{\sigma n} = \bar{\boldsymbol{t}}$
$\check{c}$	$\pi_\mathrm{d} = \operatorname{div}(\boldsymbol{\xi}_\mathrm{d})$	$\boldsymbol{\xi}_\mathrm{d} \cdot \boldsymbol{n} = \bar{\Xi}_\mathrm{d}$

9.2.3 Free energy density and dissipation inequality

The considered free energy density includes an elastic W_e, a configurational W_0, an interface W_Γ, and a penalty W_χ energy contribution in an additive way

$$W = W_\mathrm{e} + W_0 + W_\Gamma + W_\chi. \tag{9.8}$$

The dissipation inequality will be regarded in the reference placement, therefore, the contributions to the free energy density are described with respect to this placement. No plastic deformation is considered within this framework, thus, the elastic free energy density is given by

$$W_\mathrm{e} = \frac{1}{2}\boldsymbol{E} \cdot \mathbb{C}[\boldsymbol{E}]. \tag{9.9}$$

Here, $\boldsymbol{E} = (\boldsymbol{F}^\mathsf{T}\boldsymbol{F} - \boldsymbol{I})/2$ corresponds to the Green strain tensor in the reference placement. A double-well curve is used for the configurational free energy. Following the approach by Ubachs et al. (2004), such a double-well curve can be fitted by

$$W_0 = g_1\, c + g_2\,(1-c) + g_3\, c\, \ln(c) + g_4\,(1-c)\, \ln(1-c) + g_5\, c\,(1-c). \tag{9.10}$$

This simplification can improve the numerical solution procedure (e.g., Ubachs et al. (2004)). The values of g_1 to g_5 are listed in Tab. 9.1. The penalty free energy is given by

$$W_\chi = \frac{H_\chi}{2}(\check{c} - c)^2 \tag{9.11}$$

and a quadratic free energy density for the interface is considered

$$W_\Gamma = \frac{k_\mathrm{d}}{2} \mathrm{Grad}\,(\check{c}) \cdot \mathrm{Grad}\,(\check{c})\,, \tag{9.12}$$

where k_d stands for the gradient energy coefficient. The entropy flux is given as

$$\boldsymbol{q}_\mathrm{s}^\star = \frac{\boldsymbol{q}^\star - \mu_\mathrm{d} \boldsymbol{j}^\star}{T}, \tag{9.13}$$

following the classical irreversible thermodynamics (e.g., Groot and Mazur (1984)). Reconsidering the balance of internal energy (Eq. (4.28)) and the balance of entropy (Eq. (4.29)), as emphasized in Section 4.4, the dissipation inequality reads (e.g., Müller (1967))

$$\mathcal{D}_\mathrm{tot} = \int_{\mathcal{V}_0} \boldsymbol{S} \cdot \dot{\boldsymbol{E}} + \pi_\mathrm{d}^\star \dot{\check{c}} + \boldsymbol{\xi}_\mathrm{d}^\star \cdot \mathrm{Grad}\,(\dot{\check{c}}) - \mathrm{Div}\,(\tilde{\mu}_\mathrm{d} \tilde{\boldsymbol{j}}) - \dot{W} \mathrm{d}V \geq 0. \tag{9.14}$$

Here, $\pi_\mathrm{d}^\star = J \pi_\mathrm{d}$ and $\boldsymbol{\xi}_\mathrm{d}^\star = J \boldsymbol{F}^{-1} \boldsymbol{\xi}_\mathrm{d}$ correspond to π_d and $\boldsymbol{\xi}_\mathrm{d}$ in the reference placement, respectively. Further, $\boldsymbol{S}$ stands for the second Piola-Kirchhoff stress ($\boldsymbol{S} = J \boldsymbol{F}^{-1} \boldsymbol{\sigma} \boldsymbol{F}^{-\mathsf{T}}$) with respect to the reference placement and $\tilde{\mu}_\mathrm{d} = \varrho_0 \mu_\mathrm{d}$ such that $\tilde{\mu}_\mathrm{d} \tilde{\boldsymbol{j}} = \mu_\mathrm{d} \boldsymbol{j}^\star$ with μ_d as the chemical potential. In addition, the contribution of the micromorphic variable and the simplifications due to an isothermal case with a homogeneous temperature distribution were considered. Substituting the free energy density (Eq. (9.8)) into the dissipation inequality (Eq. (9.14)) and applying the material time derivative leads to

$$\begin{aligned} \mathcal{D}_\mathrm{tot} = \int_{\mathcal{V}_0} & \left(\boldsymbol{S} - \frac{\partial W_\mathrm{e}}{\partial \boldsymbol{E}} \right) \cdot \dot{\boldsymbol{E}} + \left(\boldsymbol{\xi}_\mathrm{d}^\star - \frac{\partial W_\Gamma}{\partial \mathrm{Grad}\,(\check{c})} \right) \cdot \mathrm{Grad}\,(\dot{\check{c}}) \\ & + \left(\tilde{\mu}_\mathrm{d} - \frac{\partial W_0}{\partial c} - \frac{\partial W_\mathrm{e}}{\partial c} - \frac{\partial W_\chi}{\partial c} \right) \dot{c} + \left(\pi_\mathrm{d}^\star - \frac{\partial W_\chi}{\partial \check{c}} \right) \dot{\check{c}} \\ & - \mathrm{Grad}\,(\tilde{\mu}_\mathrm{d}) \cdot \tilde{\boldsymbol{j}}\, \mathrm{d}V \geq 0. \end{aligned} \tag{9.15}$$

By substituting the specific free energy densities W_0, W_e, W_Γ, and W_χ into Eq. (9.15) as well as by assuming arbitariness of the rates $\dot{c}$, $\dot{\check{c}}$, and

$\mathrm{Grad}\,(\dot{\check{c}})$ the following relations are obtained:

$$\begin{aligned} \boldsymbol{S} &= \mathbb{C}[\boldsymbol{E}], & \pi_{\mathrm{d}}^{\star} &= H_{\chi}\,(\check{c}-c)\,, \\ \boldsymbol{\xi}_{\mathrm{d}}^{\star} &= k_{\mathrm{d}}\mathrm{Grad}\,(\check{c})\,, & \tilde{\mu}_{\mathrm{d}} &= \frac{\partial W_0}{\partial c} + \frac{1}{2}\boldsymbol{E}\frac{\partial \mathbb{C}}{\partial c}[\boldsymbol{E}] - H_{\chi}\,(\check{c}-c)\,. \end{aligned} \tag{9.16}$$

Consequently, the dissipation inequality reduces to

$$-\,\mathrm{Grad}\,(\tilde{\mu}_{\mathrm{d}}) \cdot \tilde{\boldsymbol{j}} \geq 0 \tag{9.17}$$

and can be used to motivate a thermodynamic consistent relationship for the diffusion mass flow $\tilde{\boldsymbol{j}}$.

9.3 Simulation of the coupled framework

9.3.1 Simulation setup

The results of coupled diffusional-mechanical simulations will be presented in the following sections. The implementation was realized in the infinitesimal deformation framework following closely the work of Ubachs et al. (2004). Consequently, the readability of the framework's equations is improved compared to the finite framework. The local diffusion is given by

$$\varrho \dot{c} = -\mathrm{div}\,(\boldsymbol{j}) \tag{9.18}$$

and the equilibrium equations are given in Box 9.1. Furthermore, the relations of Eq. (9.16) simplify to

$$\begin{aligned} \boldsymbol{\sigma} &= \mathbb{C}[\boldsymbol{\varepsilon}], & \pi_{\mathrm{d}} &= H_{\chi}\,(\check{c}-c)\,, \\ \boldsymbol{\xi}_{\mathrm{d}} &= k_{\mathrm{d}}\mathrm{grad}\,(\check{c})\,, & \mu_{\mathrm{d}} &= \frac{\partial W_0}{\partial c} + \frac{1}{2}\boldsymbol{\varepsilon}\frac{\partial C}{\partial c}[\boldsymbol{\varepsilon}] - H_{\chi}\,(\check{c}-c) \end{aligned} \tag{9.19}$$

in the infinitesimal deformation framework. Following the approach by Ubachs et al. (2004), the diffusion mass flow $\boldsymbol{j}$ is modeled by

$$\boldsymbol{j} = -\varrho \boldsymbol{M}_{\mathrm{d}}\mathrm{grad}\,(\mu_{\mathrm{d}}) \tag{9.20}$$

and, thus, fulfills the reduced dissipation inequality (Eq. (9.17)). Here, $\boldsymbol{M}_\mathrm{d}$ corresponds to the mobility tensor. The necessary parameters are listed in Tab. 9.1 and the relevant equations are summarized in Box 9.2.

Box 9.2: Summary of the equilibrium equations for the Cahn-Hilliard model coupled with the balance equation of linear momentum.

Field variable	Equilibrium equation
c	$\varrho\dot{c} = \mathrm{div}\left(\varrho \boldsymbol{M}_\mathrm{d}\mathrm{grad}\left(\frac{\partial W_0}{\partial c} - H_\chi(\check{c}-c) + \mu_\mathrm{e}\right)\right)$
$\check{c}$	$\check{c} - \lambda_\mathrm{d}^2\mathrm{div}\left(\mathrm{grad}\left(\check{c}\right)\right) = c$
$\boldsymbol{u}$	$\mathrm{div}\left(\boldsymbol{\sigma}\right) = \mathbf{0}$
μ_e	$\mu_\mathrm{e} = \frac{1}{2}\varepsilon \cdot \frac{\partial \mathbb{C}}{\partial c}[\varepsilon]$

Here, an additional degree-of-freedom was introduced (μ_e) to avoid the third order partial derivative of ε with respect to $\boldsymbol{x}$. This allows the use of standard quadratical shape functions of a 20-node hexahedral element for the spatial discretization (compare Section 5.2.2). Further, following Ubachs et al. (2004), the penalty parameter H_χ can be interpreted as an interface tension coefficient and $\lambda_\mathrm{d} = \sqrt{k_\mathrm{d}/H_\chi}$ as the internal length. The diffusion equation and the linear momentum were solved fully coupled, contrary to the work of Ubachs et al. (2004) where the two equations were solved iteratively. The Neumann boundary conditions $\mathrm{grad}\left(c\right) \cdot \boldsymbol{n} = 0$ and $\mathrm{grad}\left(\check{c}\right) \cdot \boldsymbol{n} = 0$ were applied within all simulations. An implicit Euler scheme was used and the condition for the equilibrium state of the simulation is given by

$$|\max\left(\mu_\mathrm{d}(\boldsymbol{x}_1)\right) - \min\left(\mu_\mathrm{d}(\boldsymbol{x}_2)\right)| \leq \epsilon\,|\min\left(\mu_\mathrm{d}(\boldsymbol{x}_1)\right)| \tag{9.21}$$

with $\epsilon \leq 2.5 \cdot 10^{-3}$ and $\forall \boldsymbol{x}_1,\ \boldsymbol{x}_2 \in \mathcal{V}$. Thus, the equilibrium state is reached when the chemical potential is homogeneous throughout the system. A concentration dependency of the density ϱ and the mobility tensor $\boldsymbol{M}_\mathrm{d}$ was assumed and modeled by an harmonic and arithmetic mean, respectively. Further, the stiffness tensor is homogenized with the lower Hashing-Shtrikman bound by assuming spherical inclusions with isotropic properties (Gross and Seelig, 2011). The residual vectors and the element stiffness matrices are given in Appendix A.4.

9.3.2 Simulation results for the Cahn-Hilliard diffusion

The diffusion process without the effects of the elastically stored energy will be considered in this section, thus, $\boldsymbol{\varepsilon} = \mathbf{0}$. To investigate the influences of the parameters H_χ and λ_d on the behavior of the interface, one-dimensional simulations were performed. The effects of H_χ and λ_d on the interface are illustrated in Fig. 9.1 a) and b), respectively. Here, the equilibrium concentration distribution is shown, which is represented by a two phase material containing an interface. An increase/decrease of λ_d and H_χ leads to a wider/narrower interface. Despite the change of the interface, the parameters affect the diffusion velocity. This is

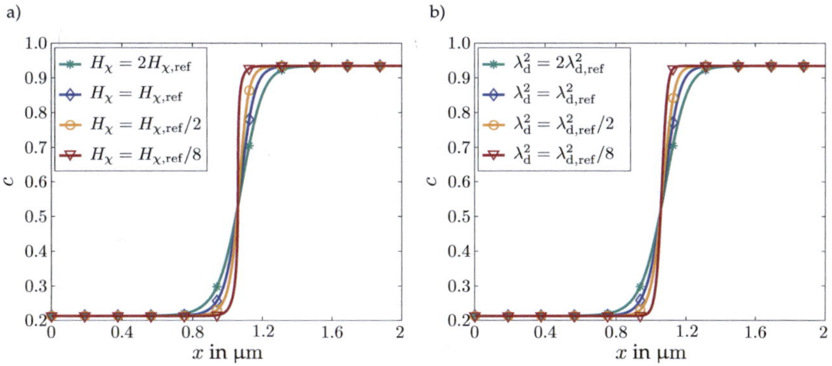

Figure 9.1: Interface in the equilibrium state for different H_χ and λ_d values depicted in a) and b), respectively. In a) $\lambda_\mathrm{d} = 2\lambda_{\mathrm{d,ref}}$ is kept constant while changing H_χ and vice versa in b) with $H_\chi = 2H_\chi$.

clearly visible by comparing the two-dimensional simulation results of the microstructure evolution for different k_d values, depicted in Fig. 9.2, where one pixel corresponds to one element of the finite element simulation. Here, the mass concentration c for three different k_d values ($k_\mathrm{d} = 4k_\mathrm{d,ref}$, $k_\mathrm{d} = k_\mathrm{d,ref}$, and $k_\mathrm{d} = k_\mathrm{d,ref}/4$, sorted in columns) at four specific times ($t_1 < t_2 < t_3 < t_4$, sorted in rows) are shown. For the k_d-modification, H_χ and λ_d^2 were changed equally. By considering the column on the left ($k_\mathrm{d} = 4k_\mathrm{d,ref}$), the lamellar coarsening process with increasing time t is visible. Approximately only half of the lamellas are present at t_4 compared to the beginning at t_1. The shrinkage process of the lamellas is visible by comparing the times t_3 and t_4 of the simulation with $k_\mathrm{d} = k_\mathrm{d,ref}$ (middle column). The relevant lamellas are marked by yellow arrows. All three simulations start with the same concentration distribution (at t_1), however, the diffusion process is slowed down with decreasing k_d (compare left column with right column for $t > t_1$).

Table 9.1: Input parameters for the Cahn-Hilliard diffusion model coupled with the balance equation of linear momentum.

General		
Description	Parameter	Value
Penalty parameter	H_χ	$8 \cdot 10^3\,\mathrm{MPa}$
Gradient energy coefficient	k_d	$2.5 \cdot 10^{-6}\,\mathrm{N}$
Internal length	$\lambda_\mathrm{d} = \sqrt{k_\mathrm{d}/H_\chi}$	$17.7 \cdot 10^{-9}\,\mathrm{m}$
Coefficients for the configurational free energy	g_1	$-1.352 \cdot 10^9\,\mathrm{Jm}^{-3}$
	g_2	$-1.359 \cdot 10^9\,\mathrm{Jm}^{-3}$
	g_3	$1.199 \cdot 10^9\,\mathrm{Jm}^{-3}$
	g_4	$0.453 \cdot 10^9\,\mathrm{Jm}^{-3}$
	g_5	$2.009 \cdot 10^9\,\mathrm{Jm}^{-3}$
Phase 1 (c)		
Description	Parameter	Value
Mass density	ϱ_1	$8 \cdot 10^3\,\mathrm{kgm}^{-3}$
Shear modulus	μ_1	$40.4 \cdot 10^3\,\mathrm{MPa}$
Poisson's ratio	ν_1	0.3
Isotropic mobility tensor ($\boldsymbol{M}_{\mathrm{d}1} = M_{\mathrm{d}1}\boldsymbol{I}$)	$M_{\mathrm{d}1}$	$10^{-25}\,\mathrm{m}^5\mathrm{J}^{-1}\mathrm{s}^{-1}$
Phase 2 ($1-c$)		
Description	Parameter	Value
Mass density	ϱ_2	$16 \cdot 10^3\,\mathrm{kgm}^{-3}$
Shear modulus	μ_2	$80.8 \cdot 10^3\,\mathrm{MPa}$
Poisson's ratio	ν_2	0.3
Isotropic mobility tensor ($\boldsymbol{M}_{\mathrm{d}2} = M_{\mathrm{d}2}\boldsymbol{I}$)	$M_{\mathrm{d}2}$	$10^{-25}\,\mathrm{m}^5\mathrm{J}^{-1}\mathrm{s}^{-1}$
Simulation setup for Section 9.3.3		
Description	Parameter	Value
Edge length of the microstructure	l_d	$32 \cdot 10^{-6}\,\mathrm{m}$
Applied external displacement	$\bar{u}$	$1.28 \cdot 10^{-6}\,\mathrm{m}$

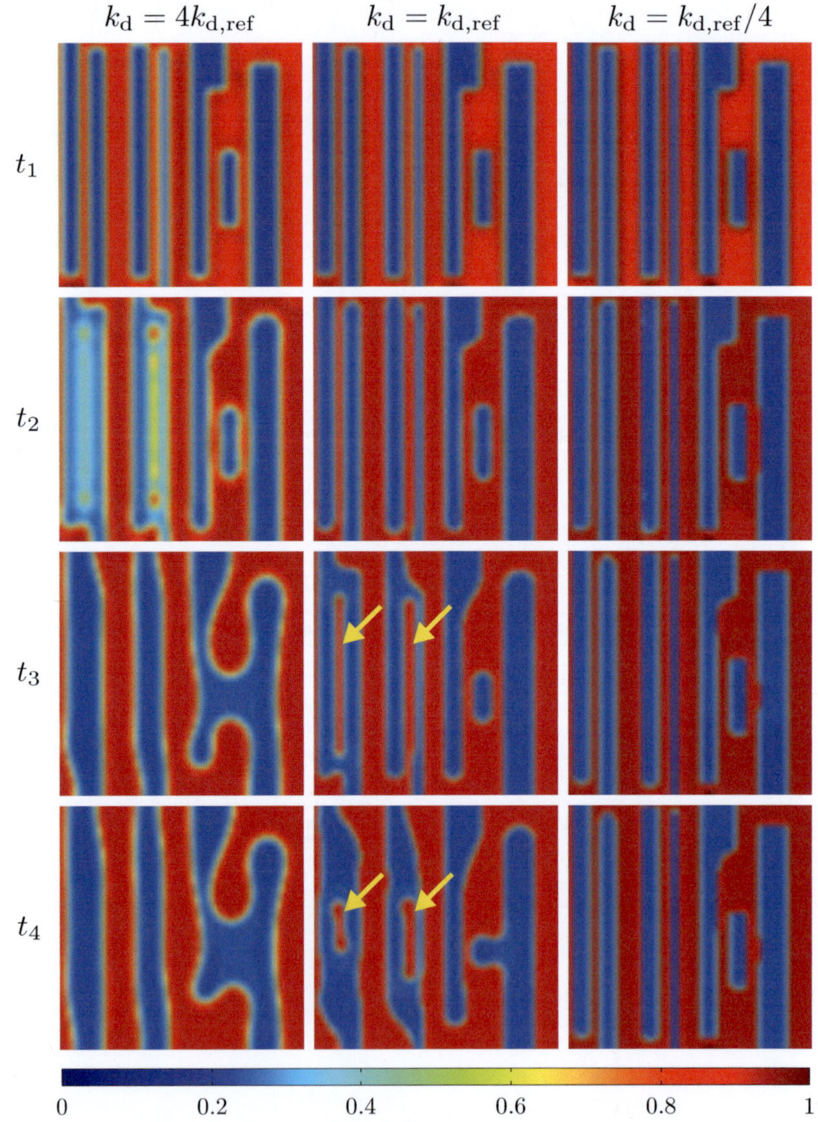

Figure 9.2: Simulation results of the Cahn-Hilliard diffusion without external load for different k_d-values (sorted in columns) at specific simulation times (sorted in rows). An increase of k_d leads to an increased diffusion velocity. A termination migration coarsening process is marked by yellow arrows.

9.3.3 Influence of an inhomogeneous strain distribution to coarsening processes

The diffusion velocity depends on the distribution of the chemical potential $\mu_\mathrm{d} = \partial W/\partial c$. As shown in the previous section, the diffusion process is slowed down with a decreasing k_d or, more specifically, with a decreasing impact of the surface free energy density W_Γ. Consequently, in such a case, other contributions to the chemical potential may control the diffusion velocity. A possible impact on the diffusion process may arise due to the elastically stored energy W_e through an inhomogeneous distribution of $\mu_\mathrm{e} = \partial W_\mathrm{e}/\partial c$. Such an inhomogeneous distribution exists for materials under structural applications with e.g., a fine lamellar microstructure within colonies/grains, where the lamellar orientation of each colony differs. To investigate the impact of an inhomogeneously distributed μ_e on the diffusion process, the microstructure as depicted in Fig. 9.3 is considered. Here, two colonies of fine lamellas with different orientations are shown (lamellas within colony ① and colony ② are parallel and perpendicular to the applied displacement $\bar{u}$, respectively). The size of the microstructure and the extent of the applied displacement are given in Tab. 9.1. The balance equation of linear momentum and the Cahn-Hilliard diffusion are solved fully coupled with the relevant equations summarized in Box 9.1. The used material parameters are listed in Tab. 9.1. Symmetric displacement boundary conditions are applied on the left and on the bottom (Fig. 9.3). The simulation is divided into three time steps: In the first step, with a total time of 20 s, the initial concentration value is smoothed without any applied external displacement. Afterwards, in the second step, the displacement $\bar{u}$ is applied with a constant rate over a total time of 20 s. Finally, the applied displacement $\bar{u}$ is kept constant over the third step with a total time of $\approx 8 \cdot 10^5$ s. Due to the negligible total times of the first two steps compared to the total time of the third step, one can conclude that the relevant diffusion takes place within the third step.

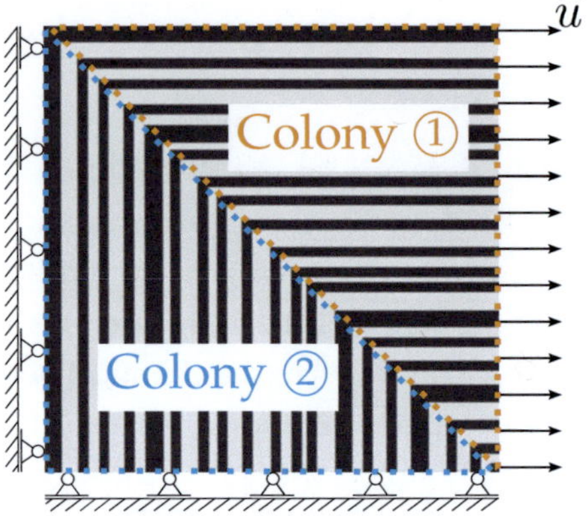

Figure 9.3: Schematical representation of the considered microstructure with the displacement boundary conditions. The lamellas of the colonies are oriented parallel and perpendicular to the external displacement $\bar{u}$, respectively.

The simulation results are shown in Fig. 9.4 for specific times (sorted in rows). In the first column, the mass concentration c is depicted. At the start of the simulation (t_1), the shape of the lamellas within each colony is still intact. After the external displacement $\bar{u}$ was applied, one can observe an coarsening at the colony boundary (domain marked by the dashed line at t_2). This coarsening intensifies with time (compare t_3 with t_2). The coarsened domain does not grow symmetrically, but rather into colony ②. To highlight this unbalanced growth, the initial colony boundary is depicted with the dotted line at t_3. Further, the amount of lamellas within colony ② is less than the amount of lamellas within colony ①. This difference is even more pronounced at t_4.
For sake of comparison, the same simulation was also conducted without an external displacement. The results are shown in the right column of Fig. 9.4. The diffusion process is slowed down compared to the diffusion process with an external displacement. In addition, the coarsening in both colonies is symmetrical.

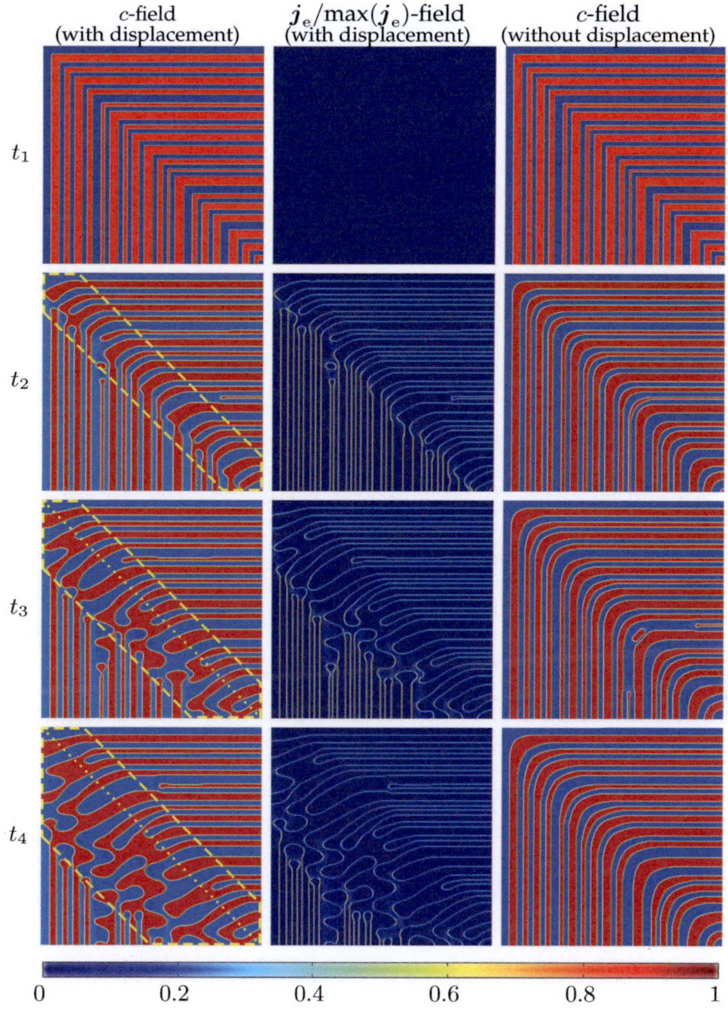

Figure 9.4: Simulation results at four specific time steps. Left column: Mass concentration distribution of the simulation with an external displacement. An increased coarsening rate is observed at the colony boundary, marked by the dashed line at t_2. This regime growth unsymmetrically, as indicated by the dotted lines at t_3 and t_4. Middle column: Absolut value of $\boldsymbol{j}_e$ normalized by its maximum value of all four illustrated time steps. The diffusion flux $\boldsymbol{j}_e$ is greater at lamellas oriented perpendicular to $\bar{u}$ compared to lamellas oriented parallel to $\bar{u}$. Right column: Mass concentration distribution of the simulation without an external displacement, for sake of comparison.

9.4 Discussion

The one-dimensional simulations described in Section 9.3.2 show the influence of the two parameters λ_d and H_χ on the interface in the equilibrium state. Both parameters affect the interface in a similar manner: an increase/reduction of λ_d and H_χ leads to a wider/narrower interface, shown in Fig. 9.1. This is in agreement with the findings of Ubachs et al. (2004), who investigated the impact of the parameters in detail. The length λ_d influences the width of the interface where the surface tension develops and H_χ controls the intensity of the surface tension (Ubachs et al., 2004). Moreover, the gradient energy coefficient $k_\mathrm{d} = \lambda_\mathrm{d}^2 H_\chi$ which controls the impact of the free energy density of the interface, is also dependent on these two parameters. A reduction of the gradient energy coefficient slows down the diffusion process (shown in Fig. 9.2). This can be rationalized by the fact that a decreasing k_d lowers the impact of the free energy density of the interface. Consequently, the diffusion flux initiated by an inhomogeneous mass concentration distribution is reduced. A termination migration coarsening process is shown in Fig. 9.2 for the simulation with $k_\mathrm{d} = k_\mathrm{d,ref}$ (middle column). Here, the reduction of the lamella length between times t_3 and t_4 is clearly visible for the two marked lamellas (marked by the yellow arrows). Such a termination migration coarsening process was also observed for e.g., Ti-17 alloys (Xu et al., 2015) and can be attributed to the free energy density of the interface. As motivated in Section 9.3.3, a decreasing free energy density of the surface W_Γ enhances the impact of other contributions to the chemical potential. Therefore, the effect of an inhomogeneous distribution of $\mu_\mathrm{e} = \partial W_\mathrm{e}/\partial c$ was investigated. The considered microstructure is shown in Fig. 9.3 and the evolution of the mass concentration is depicted in the left column in Fig. 9.4. It was shown that the coarsening of the lamellas is initiated at the colony boundary (Fig. 9.4, left column at t_2). This can be rationalized by considering the free energy density of the surface W_Γ which penalizes strong inhomogeneous mass concentrations. Compared to the diffusion process without an external displacement (right column), the domain with the increased coarsening grows not symmetrically, but rather into colony ②. This unsymmetrical growth is highlighted by illustrating the initial colony boundary with the dotted line in Fig. 9.4 at t_3. It can be explained by considering the flux

$$\boldsymbol{j}_{\mathrm{e}} = -\varrho \boldsymbol{M}_{\mathrm{d}} \operatorname{grad}(\mu_{\mathrm{e}}) \tag{9.22}$$

which arises due to the inhomogeneous distribution of μ_{e}. The absolute value of $\boldsymbol{j}_{\mathrm{e}}$ normalized by its maximum value of all four illustrated time steps is depicted in Fig. 9.4 in the middle column. At t_1, the external displacement is not applied yet, therefore, the flux $\boldsymbol{j}_{\mathrm{e}}$ is still zero within both colonies, whereas for times $t > t_1$ the flux $\boldsymbol{j}_{\mathrm{e}}$ is visible. Compared to the simulation without an external displacement, this additional flux leads to the increased diffusion process (compare the diffusion velocity shown in the left column with that in the right column in Fig. 9.4). Further, as expected due to the different lamellar orientations between both colonies, the flux in colony ② exceeds the one in colony ①, especially where the lamellas are oriented perpendicularly to the external displacement. The increased flux in colony ② leads to the unsymmetrical growth of the domain with increased coarsening. In addition, it causes the faster vanishing of lamellas within colony ② compared to colony ①. This can be seen at t_3 and even more pronounced at t_4, where the amount of lamellas within colony ② is less compared to colony ①. As discussed in Section 3.3, the interface of the lamella can act as an obstacle to hinder the movement of dislocations and, thus, the amount of lamellar interfaces affects the resistance against plastic flow. Therefore, to analyze the amount of lamellar interfaces within each grain, an edge detection algorithm was used. The detected edges for time steps t_2, t_3, and t_4 are depicted in Fig. 9.5 a) by the lines. Simply counting the interface points in each colony gives a first approximation of the total length of interfaces $l_{①}$ and $l_{②}$ within colony ① and ②, respectively. The ratio of both lengths is shown in Fig. 9.5 b) over the simulation time. It is clearly visible that the total interface length in colony ① exceeds the one in colony ②. Consequently, the resistance against plastic flow should be higher in colony ① compared to colony ②, since a higher number of interfaces can act as dislocation obstacles.

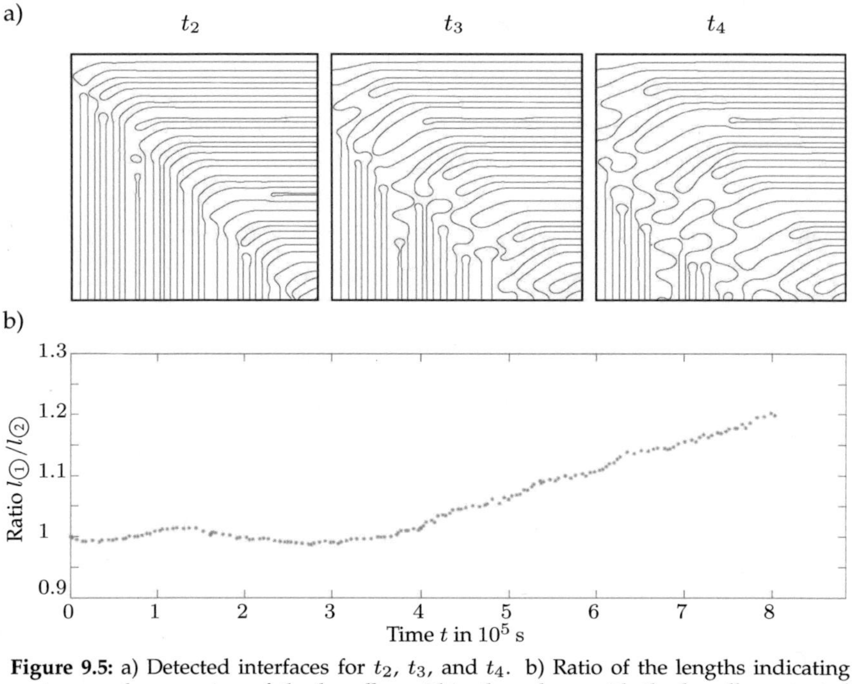

Figure 9.5: a) Detected interfaces for t_2, t_3, and t_4. b) Ratio of the lengths indicating an increased coarsening of the lamellas within the colony with the lamellas oriented perpendicular to $\bar{u}$.

These findings are interesting, especially to describe the creep mechanism for materials under structural applications and at high temperatures which possess an inhomogeneous distribution of μ_e. This is the case, e.g., for the Fe-61Al intermetallic compound, which exhibits a fine lamellar microstructure within colonies, where the lamellar orientation of each colony differs. In the beginning, the lamellar structure of each colony is still intact and the lamellas terminate closely to the colony boundaries. However, the simulation results suggest that due to an applied external load, the diffusion process is intensified and the fine lamellar structure gets lost at the colony boundaries. Consequently, this would lead to a loss in the resistance against plastic flow, which was experimentally observed for the Fe-61Al intermetallic compound by Schmitt et al. (2017). Further, the simulation results show that the lamellar coarsening is more pronounced for colonies with the lamellar orientation perpendicular to the external load.

9.5 Summary and conclusion

To account for the microstructural change over time, the proposed simulation model couples the balance equation of linear momentum with the Cahn-Hilliard diffusion through the elastically stored energy. First, the diffusion process without an external displacement was analyzed. It was shown that the parameters λ_{d} and H_{χ} not only influence the interface width, but also affect the velocity of the diffusion process through k_{d}. The main focus of the work lay on the investigation of the effect of an inhomogeneous strain distribution on the coarsening process. Here, the lamellar coarsening of two colonies was considered, which contain fine lamellas with different lamellar orientations, respectively. By considering an applied external displacement within the coupled simulations, the main findings are:

- An increased lamellar coarsening at the colony boundary was observed.
- The lamellar coarsening at the colony boundary is more pronounced in colony ②. This unsymmetrical behavior is induced by the increased mass flux in colony ② compared to colony ①, which results from the different lamellar orientations.

- A colony with the lamellar orientation perpendicular to the external load shows an increased mass diffusion flux compared to a colony with lamellas oriented parallel to the external load.
- The diffusion process velocity is increased by an external load.

As discussed in Section 3.3, the interface of a lamella can hinder the movement of dislocations. Consequently, the described microstructural change can strongly influence the composite's resistance against plastic deformation and, thus, the creep resistance. Further, the lamellar coarsening process is often identified experimentally without an external load only. However, the simulations reveal an increased velocity of the diffusion process due to the external load, thus, this should also be considered by the experimental investigations of the coarsening process.

Mesh convergency was investigated for the simulation results depicted in Fig. 9.1. Nevertheless, mesh convergency of the remaining simulations should be considered within further studies. Furthermore, the simulation model incorporates elastic deformations only. Therefore, the model should be expanded to consider plastic deformations. In addition, the considered microstructure is rather small, consequently, it would be desirable to analyze the diffusion process within a larger microstructure. Preferably, this simulation should be based on an experimentally observed microstructure for sake of validation.

Chapter 10

Summary and conclusion

Within this work, several open questions of the thermo-mechanical material behavior at high temperatures were addressed, with the focus on directionally solidified eutectics. For this purpose, three physical models were developed, namely the single-crystal plasticity model, the gradient crystal plasticity model, and the coupled Cahn-Hilliard diffusion model. The relevant equilibrium equations were derived for each model by following a general micromorphic approach (Section 4.5). A summary of the three models together with specific constitutive assumptions are given in Tab. 10.1, whereas the boundary conditions are not given for sake of readability. For all simulation models, the validation by a direct comparison to experimental observations has to be emphasized. The single-crystal plasticity model was used to describe the creep behavior of the DS NiAl-9Mo and to allow an insight into the composite effect. It is noteworthy that the model is only based on measurable quantities or parameters restricted to a physically reasonable range. Consequently, it is also applicable to other DS eutectics, having the same matrix but different reinforcements (e.g., NiAl-31Cr-3Mo, shown in Section 7.5). To describe the plastic behavior of the reinforcing phases, the derived size-dependent transition model which relates the strength of a material with the dislocation density is crucial. It considers the transition from the defect-free behavior to a bulk material behavior and fulfills multiple relevant experimental observations.

Table 10.1: Overview of the equilibrium equations and specific constitutive assumptions of the three developed simulation models.

	Equilibrium condition $\forall \boldsymbol{x} \in \mathcal{V}$	Jump condition on Γ	Constitutive assumption	
Single-crystal plasticity (Chapter 7)	$\mathbf{0} = \operatorname{div}(\boldsymbol{\sigma})$		$\boldsymbol{F} = \boldsymbol{F}_\mathrm{e}\boldsymbol{F}_\mathrm{p}$ $\dot{\gamma} = \sum_{\alpha=1}^{N_\mathrm{GS}} \dot{\gamma}_\alpha$ $\tau^\mathrm{F} = \frac{\tau_\infty}{d\sqrt{\rho}+1} + \alpha\mu b\sqrt{\rho}$ $\boldsymbol{\sigma} = J^{-1}\boldsymbol{F}_\mathrm{e}\boldsymbol{S}_\mathrm{e}\boldsymbol{F}_\mathrm{e}^\mathsf{T}$	$\boldsymbol{L}_\mathrm{p} = \sum_{\alpha=1}^{N_\mathrm{GS}} \dot{\gamma}_\alpha \tilde{\boldsymbol{M}}_\alpha$ $\dot{\gamma}_\alpha = \dot{\gamma}_0 \operatorname{sgn}(\tau_\alpha) \exp\left(-\frac{Q_\mathrm{C}}{RT}\right)\left(\frac{\lvert\tau_\alpha\rvert}{\tau_\alpha^\mathrm{F}}\right)^m$ $\dot{\rho} = \left(k_1\sqrt{\rho} - k_2\rho\right)\dot{\gamma}$
Gradient crystal plasticity (Chapter 8)	$\mathbf{0} = \operatorname{div}(\boldsymbol{\sigma})$ $\pi = \operatorname{div}(\boldsymbol{\xi})$	$[\![\boldsymbol{\sigma}]\!]\boldsymbol{n}_\Gamma = \mathbf{0}$ $\Xi = [\![\boldsymbol{\xi}]\!] \cdot \boldsymbol{n}_\Gamma$	$\boldsymbol{\varepsilon} = \boldsymbol{\varepsilon}_\mathrm{e} + \boldsymbol{\varepsilon}_\mathrm{p}$ $\dot{\gamma} = \sum_{\alpha=1}^{N_\mathrm{GS}} \dot{\gamma}_\alpha$ $\tau^\mathrm{F} = \frac{\tau_\infty}{d\sqrt{\rho_\mathrm{ssd}}+1} + \alpha\mu b\sqrt{\rho_\mathrm{ssd}}$ $W_\mathrm{e} = \frac{1}{2}\boldsymbol{\varepsilon}_\mathrm{e} \cdot \mathbb{C}[\boldsymbol{\varepsilon}_\mathrm{e}]$ $W_\chi = \frac{H_\chi}{2}(\gamma_\mathrm{acc} - \check{\gamma})^2$ $0 = (\Xi - k_\Gamma - \Xi^\mathrm{F})\dot{\check{\gamma}} = f_\Gamma \dot{\check{\gamma}}$ $\Xi^\mathrm{F} = k_\sigma \langle -\boldsymbol{\sigma} \cdot (\boldsymbol{n}_\Gamma \otimes \boldsymbol{n}_\Gamma)\rangle$	$\boldsymbol{\varepsilon}_\mathrm{p} = \sum_{\alpha=1}^{N_\mathrm{GS}} \gamma_\alpha \boldsymbol{M}_\alpha^\mathrm{S}$ $\dot{\gamma}_\alpha = \dot{\gamma}_0 \operatorname{sgn}(\tau_\alpha)\left(\frac{\langle\lvert\tau_\alpha\rvert - \check{p}\rangle}{\tau_\alpha^\mathrm{F}}\right)^m$ $\dot{\rho}_\mathrm{ssd} = k_1\sqrt{\rho_\mathrm{ssd}}\dot{\check{\gamma}}$ $W_\nabla = \frac{1}{2}k_\nabla \operatorname{grad}(\check{\gamma}) \cdot \operatorname{grad}(\check{\gamma})$ $W_\Gamma = k_\Gamma \check{\gamma}$
Coupled Cahn-Hilliard diffusion (Chapter 9)	$\mathbf{0} = \operatorname{div}(\boldsymbol{\sigma})$ $\varrho\dot{c} = \operatorname{div}(\varrho \boldsymbol{M}_\mathrm{d} \operatorname{grad}(\mu_\mathrm{d}))$ $\check{c} - \lambda_\mathrm{d}^2 \operatorname{div}(\operatorname{grad}(\check{c})) = c$ $\mu_\mathrm{e} = \frac{1}{2}\boldsymbol{\varepsilon} \cdot \frac{\partial \mathbb{C}}{\partial c}[\boldsymbol{\varepsilon}]$		$\boldsymbol{\varepsilon} = \boldsymbol{\varepsilon}_\mathrm{e}$ $\boldsymbol{j} = -\varrho \boldsymbol{M}_\mathrm{d} \operatorname{grad}(\mu_\mathrm{d})$ $W_\chi = \frac{H_\chi}{2}(\check{c} - c)^2$ $W_0 = g_1 c + g_2(1-c) + g_3 c \ln(c) + g_4(1-c)\ln(1-c) + g_5 c(1-c)$	$\mu_\mathrm{d} = \frac{\partial W_0}{\partial c} + \frac{\partial W_\mathrm{e}}{\partial c} + \frac{\partial W_\chi}{\partial c}$ $W_\mathrm{e} = \frac{1}{2}\boldsymbol{\varepsilon} \cdot \mathbb{C}[\boldsymbol{\varepsilon}]$ $W_\Gamma = \frac{k_\mathrm{d}}{2} \operatorname{Grad}(\check{c}) \cdot \operatorname{Grad}(\check{c})$

Furthermore, it is in agreement with the transition model proposed by El-Awady (2015), who analyzed over 200 discrete dislocation dynamics simulations supported by micro-pillar experiments of several materials. Consequently, the transition model described in this work is not only applicable to the reinforcing phases of DS materials, but also valid for other materials. Finally, it is valid in an extended range of dislocation densities compared to the model of El-Awady (2015). The single-crystal plasticity model was able to describe the creep behavior of the DS NiAl-9Mo at various creep conditions (changed temperature and applied load) as well as for different withdrawal rates of the directional solidification process. Further, the impact of an increased fiber content on the composite's behavior was analyzed.
To investigate the effects of dislocation pile-ups at the fiber-matrix interface as well as the slip transfer/activation, a gradient crystal plasticity model was used. Here, the IGFEM approach with enriched shape functions was applied and validated in the context of gradient plasticity. This approach is well-suited for moving sharp interfaces. The gradient crystal plasticity model incorporates not only the impact of the geometrically necessary dislocations due to the inhomogeneous plastic deformation, but also a stress dependent interface flow rule. In addition, the essentially defect-free state of the fibers could be considered. Both, the single-crystal plasticity model as well as the gradient crystal plasticity model were compared to experimental measurements and one can conclude that both models show a good agreement with the experiments. The simulations reveal that the fiber phase and the fiber-matrix interface control the composite's behavior.
The third simulation model considers the coupling between the balance equation of linear momentum and the Cahn-Hilliard diffusion. Special emphasis was placed on the diffusion process of a colony based microstructure with fine lamellas within individual colonies. Here, the effect of an inhomogeneous strain distribution on the lamellar coarsening was investigated. Furthermore, the consequences of this diffusion

process on the creep resistance was discussed. In summary, the highlights of this work are:

- Size-dependent transition model to describe the strength of a material as a function of the dislocation density. The strength decreases starting from the limit of the theoretical strength in the case of a defect-free material and reaches the normal square root Taylor hardening for high densities. Further, it fulfills important experimental findings of the material strength, such as (i) the size-independency for a material with low or high dislocation densities, (ii) the coupling between the dislocation density and the size parameter, resulting in a non-constant exponent n, and (iii) the reduction of the size-dependent regime with increasing dislocation density.
- Single-crystal plasticity model for DS eutectics having a NiAl based matrix but different reinforcements, whereas the research focus was set on the DS NiAl-9Mo. The creep behavior of the composite could be well described, even for (i) a change of the applied stress, (ii) a change of the temperature, and (iii) a change of the fiber diameter. The simulations allow an insight into the composite effect as well as the prediction of the influences of an increased fiber content and a further increased/decreased withdrawal rate on the creep behavior.
- The gradient crystal plasticity model extends the local model by incorporating the effects of geometrically necessary dislocations. Further, it considers the initial elastic behavior of the fibers due to their essentially defect-free state. The pile-up stress at the fiber-matrix interface leads to the transfer/activation of slip, initiating the plastic deformation of the fibers.
- Stress dependent interface flow rule motivated by several investigations from literature. It considers the normal stress on the grain boundary and was validated at high temperatures.

- Cahn-Hilliard diffusion model coupled with the balance equation of linear momentum. The highest coarsening rate was observed at the boundary of two colonies, which contain fine lamellas with different lamellar orientation, respectively. In addition, the colony with the lamellas oriented perpendicular to the external load showed an increased lamellar coarsening velocity.

Even though multiple relevant open questions of the thermo-mechanical material behavior at high temperatures have been addressed within this work, further investigations are necessary. The dislocation density evolution as a function of the creep strain should be investigated experimentally to enable the validation of the proposed dislocation evolution by a direct comparison. In order to analyze the interfaces' behavior at high temperatures in more detail, further comparisons between experiments and simulations at several temperatures and various fiber spacings/diameters are necessary. Furthermore, an improved physical insight may arise by considering all plastic slips individually. Finally, the coupled Cahn-Hilliard diffusion model should be applied to an experimentally scanned microstructure.

Appendix A

Appendix

A.1 Equilibrium equations including micromorphic variables

The balance of virtual power of internal and external forces, thus $\delta\mathcal{P}_{\mathrm{int}} = \delta\mathcal{P}_{\mathrm{ext}}$, reads

$$\begin{aligned}&\int\limits_{\mathcal{V}} \boldsymbol{\sigma} \cdot \delta\boldsymbol{L} + \boldsymbol{\xi}_{\mathrm{G}} \cdot \operatorname{grad}\left(\delta\dot{\check{\phi}}\right) + \pi_{\mathrm{G}}\delta\dot{\check{\phi}}\mathrm{d}v + \int\limits_{\Gamma} \Xi^{\mathrm{s}}_{\mathrm{G}}\langle\!\langle\delta\dot{\check{\phi}}\rangle\!\rangle + \Xi^{\mathrm{j}}_{\mathrm{G}}[\![\delta\dot{\check{\phi}}]\!]\mathrm{d}a\\&= \int\limits_{\mathcal{A}_{\mathrm{t}}} \bar{\boldsymbol{t}} \cdot \delta\dot{\boldsymbol{u}}\mathrm{d}a + \int\limits_{\mathcal{A}_{\Xi}} \bar{\Xi}_{\mathrm{G}}\delta\dot{\check{\phi}}\mathrm{d}a.\end{aligned} \tag{A.1}$$

Here the virtual powers $\delta\mathcal{P}_{\mathrm{int}}$ and $\delta\mathcal{P}_{\mathrm{ext}}$ given in Eq. (4.35) and Eq. (4.37) were used. The first term of Eq. (A.1) can be reformulated to

$$\begin{aligned}\int\limits_{\mathcal{V}} \boldsymbol{\sigma} \cdot \delta\boldsymbol{L}\mathrm{d}v = &- \int\limits_{\mathcal{V}} \operatorname{div}(\boldsymbol{\sigma}) \cdot \delta\dot{\boldsymbol{u}}\mathrm{d}v\\&+ \int\limits_{\mathcal{A}_{\mathrm{t}}} \boldsymbol{\sigma}\boldsymbol{n} \cdot \delta\dot{\boldsymbol{u}}\mathrm{d}a - \int\limits_{\Gamma} [\![\boldsymbol{\sigma}]\!]\boldsymbol{n}_{\Gamma} \cdot \delta\dot{\boldsymbol{u}}\mathrm{d}a,\end{aligned} \tag{A.2}$$

by using the relation

$$\operatorname{div}(\boldsymbol{\sigma}\delta\dot{\boldsymbol{u}}) = \boldsymbol{\sigma} \cdot \delta\boldsymbol{L} + \operatorname{div}(\boldsymbol{\sigma}) \cdot \delta\dot{\boldsymbol{u}}, \tag{A.3}$$

the divergence theorem

$$\int_{\mathcal{V}} \operatorname{div}(\boldsymbol{\sigma}\delta\dot{\boldsymbol{u}})\,\mathrm{d}v = \int_{\mathcal{A}_\mathrm{t}} (\boldsymbol{\sigma}\delta\dot{\boldsymbol{u}})\cdot\boldsymbol{n}\mathrm{d}a - \int_{\Gamma} [\![\boldsymbol{\sigma}\delta\dot{\boldsymbol{u}}]\!]\cdot\boldsymbol{n}_\Gamma \mathrm{d}a, \tag{A.4}$$

and the symmetry of the Cauchy stress $\boldsymbol{\sigma} = \boldsymbol{\sigma}^\mathsf{T}$. Following the same procedure, the second term of Eq. (A.1) reads

$$\begin{aligned}\int_{\mathcal{V}} \boldsymbol{\xi}_\mathrm{G}\cdot\operatorname{grad}\left(\delta\dot{\check{\phi}}\right)\mathrm{d}v = &-\int_{\mathcal{V}}\operatorname{div}(\boldsymbol{\xi}_\mathrm{G})\cdot\delta\dot{\check{\phi}}\mathrm{d}v + \int_{\mathcal{A}_\Xi}\boldsymbol{\xi}_\mathrm{G}\cdot\boldsymbol{n}\delta\dot{\check{\phi}}\mathrm{d}a\\ &-\int_{\Gamma}[\![\boldsymbol{\xi}_\mathrm{G}]\!]\cdot\boldsymbol{n}_\Gamma\langle\!\langle\delta\dot{\check{\phi}}\rangle\!\rangle + \langle\!\langle\boldsymbol{\xi}_\mathrm{G}\rangle\!\rangle\cdot\boldsymbol{n}_\Gamma[\![\delta\dot{\check{\phi}}]\!]\mathrm{d}a.\end{aligned} \tag{A.5}$$

By considering the two equations (Eq. (A.2) and Eq. (A.5)), the balance of virtual power of internal and external forces results in

$$\begin{aligned}&\int_{\mathcal{V}} -\operatorname{div}(\boldsymbol{\sigma})\cdot\delta\dot{\boldsymbol{u}} + \pi_\mathrm{G}\delta\dot{\check{\phi}} - \operatorname{div}(\boldsymbol{\xi}_\mathrm{G})\cdot\delta\dot{\check{\phi}}\mathrm{d}v + \int_{\Gamma} -[\![\boldsymbol{\sigma}]\!]\boldsymbol{n}_\Gamma\cdot\delta\dot{\boldsymbol{u}}\\ &+(\Xi^\mathrm{s}_\mathrm{G} - [\![\boldsymbol{\xi}_\mathrm{G}]\!]\cdot\boldsymbol{n}_\Gamma)\langle\!\langle\delta\dot{\check{\phi}}\rangle\!\rangle + (\Xi^\mathrm{j}_\mathrm{G} - \langle\!\langle\boldsymbol{\xi}_\mathrm{G}\rangle\!\rangle\cdot\boldsymbol{n}_\Gamma)[\![\delta\dot{\check{\phi}}]\!]\mathrm{d}a\\ &+\int_{\mathcal{A}_\mathrm{t}}(\boldsymbol{\sigma}\boldsymbol{n} - \bar{\boldsymbol{t}})\cdot\delta\dot{\boldsymbol{u}}\mathrm{d}a + \int_{\mathcal{A}_\Xi}(\boldsymbol{\xi}_\mathrm{G}\cdot\boldsymbol{n} - \bar{\Xi}_\mathrm{G})\delta\dot{\check{\phi}}\mathrm{d}a = 0.\end{aligned} \tag{A.6}$$

The arbitrariness of the two rates $\delta\dot{\boldsymbol{u}}$, $\delta\dot{\phi}$ yields the equilibrium equations summarized in Box 4.2.

A.2 Equations for the periodic boundary conditions

The periodic boundary conditions are derived for the displacement degree-of-freedom. Nevertheless, they can be applied for any desired DOF. Here, the rectangle with the dimensions L_x, L_y, and L_z in the x, y, and z direction, depicted in Fig. 5.7, is considered as the periodic unit cell. As emphasized in Section 5.3, the independent nodal displacements are u_{x0}, u_{y0}, u_{z0}, u_{x1}, u_{y1}, u_{z1}, u_{x2}, u_{y2}, and u_{x4}. Here, the first index represents the axial direction of the displacement and the second index the node number according to Fig. 5.7. By controlling the independent nodal displacements, it is possible to adjust the desired effective strain tensor (Eq. 5.3). In order to facilitate the reading of the periodic boundary conditions, the nodes are sorted by their position in the following sets: corner nodes, edges nodes, and surface nodes.

A.2.1 Coupling equation of corner nodes

The remaining displacements of the corner nodes are restricted by equations (A.7) to (A.21).

$$u_{z2} = u_{z0} + \frac{L_y}{L_z}\left(u_{y1} - u_{y0}\right) \tag{A.7}$$

$$u_{x3} = u_{x1} + u_{x2} - u_{x0} \tag{A.8}$$

$$u_{y3} = u_{y1} + u_{y2} - u_{y0} \tag{A.9}$$

$$u_{z3} = u_{z1} + \frac{L_y}{L_z}\left(u_{y1} - u_{y0}\right) \tag{A.10}$$

$$u_{y4} = u_{y0} + \frac{L_x}{L_y}\left(u_{x2} - u_{x0}\right) \tag{A.11}$$

$$u_{z4} = u_{z0} + \frac{L_x}{L_z}\left(u_{x1} - u_{x0}\right) \tag{A.12}$$

$$u_{x5} = u_{x1} + u_{x4} - u_{x0} \tag{A.13}$$

$$u_{y5} = u_{y1} + \frac{L_x}{L_y}\left(u_{x2} - u_{x0}\right) \tag{A.14}$$

$$u_{z5} = u_{z1} + \frac{L_x}{L_z}\left(u_{x1} - u_{x0}\right) \tag{A.15}$$

$$u_{x6} = u_{x4} + u_{x2} - u_{x0} \tag{A.16}$$

$$u_{y6} = u_{y2} + \frac{L_x}{L_y}(u_{x2} - u_{x0}) \tag{A.17}$$

$$u_{z6} = u_{z0} + \frac{L_y}{L_z}(u_{y1} - u_{y0}) + \frac{L_x}{L_z}(u_{x1} - u_{x0}) \tag{A.18}$$

$$u_{x7} = u_{x1} + u_{x2} + u_{x4} - 2u_{x0} \tag{A.19}$$

$$u_{y7} = u_{y1} + u_{y2} - u_{y0} + \frac{L_x}{L_y}(u_{x2} - u_{x0}) \tag{A.20}$$

$$u_{z7} = u_{z1} + \frac{L_x}{L_z}(u_{x1} - u_{x0}) + \frac{L_y}{L_z}(u_{y1} - u_{y0}) \tag{A.21}$$

A.2.2 Coupling equation of edge nodes

The set of edge nodes contains all nodes on the edge of the unit cell, except the corner nodes. This set is further divided into edge node subsets, depicted in Fig. A.2.1. Here, each subset possesses one master-edge (green edge in Fig. A.2.1) containing the master-nodes (indicated by the superscript $(\cdot)^{\mathrm{M}}$) and three slave-edges (red edges in Fig. A.2.1) containing the slave-nodes (indicated by the superscript $(\cdot)^{\mathrm{S1}}$, $(\cdot)^{\mathrm{S2}}$, $(\cdot)^{\mathrm{S3}}$). The displacement of each slave-node depends on the displacement of the corresponding master-node and the independent nodal displacements.

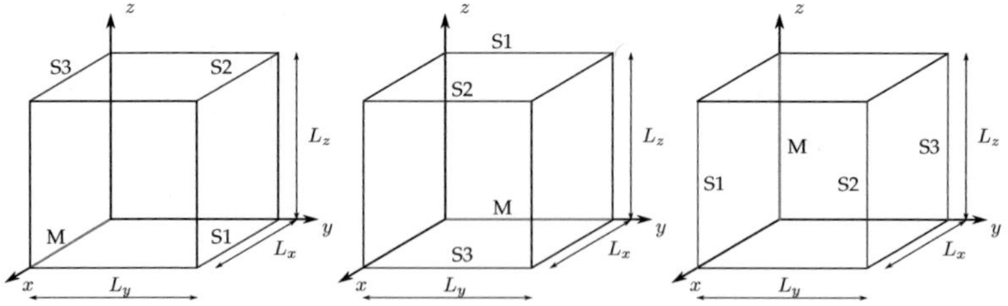

Figure A.2.1: Subsets of the edge nodes.

Edge node subset x

$$u_x^{\mathrm{S1}} = u_x^{\mathrm{M}} + u_{x2} - u_{x0} \tag{A.22}$$
$$u_y^{\mathrm{S1}} = u_y^{\mathrm{M}} + u_{y2} - u_{y0} \tag{A.23}$$
$$u_z^{\mathrm{S1}} = u_z^{\mathrm{M}} + \frac{L_y}{L_z}\left(u_{y1} - u_{y0}\right) \tag{A.24}$$

$$u_x^{\mathrm{S2}} = u_x^{\mathrm{M}} + u_{x2} + u_{x1} - 2u_{x0} \tag{A.25}$$
$$u_y^{\mathrm{S2}} = u_y^{\mathrm{M}} + u_{y2} + u_{y1} - 2u_{y0} \tag{A.26}$$
$$u_z^{\mathrm{S2}} = u_z^{\mathrm{M}} + u_{z1} - u_{z0} + \frac{L_y}{L_z}\left(u_{y1} - u_{y0}\right) \tag{A.27}$$

$$u_x^{\mathrm{S3}} = u_x^{\mathrm{M}} + u_{x1} - u_{x0} \tag{A.28}$$
$$u_y^{\mathrm{S3}} = u_y^{\mathrm{M}} + u_{y1} - u_{y0} \tag{A.29}$$
$$u_z^{\mathrm{S3}} = u_z^{\mathrm{M}} + u_{z1} - u_{z0} \tag{A.30}$$

Edge node subset y

$$u_x^{\mathrm{S1}} = u_x^{\mathrm{M}} + u_{x1} - u_{x0} \tag{A.31}$$
$$u_y^{\mathrm{S1}} = u_y^{\mathrm{M}} + u_{y1} - u_{y0} \tag{A.32}$$
$$u_z^{\mathrm{S1}} = u_z^{\mathrm{M}} + u_{z1} - u_{z0} \tag{A.33}$$

$$u_x^{\mathrm{S2}} = u_x^{\mathrm{M}} + u_{x1} + u_{x4} - 2u_{x0} \tag{A.34}$$
$$u_y^{\mathrm{S2}} = u_y^{\mathrm{M}} + u_{y1} - u_{y0} + \frac{L_x}{L_y}\left(u_{x2} - u_{x0}\right) \tag{A.35}$$
$$u_z^{\mathrm{S2}} = u_z^{\mathrm{M}} + u_{z1} - u_{z0} + \frac{L_x}{L_z}\left(u_{x1} - u_{x0}\right) \tag{A.36}$$

$$u_x^{\mathrm{S3}} = u_x^{\mathrm{M}} + u_{x4} - u_{x0} \tag{A.37}$$
$$u_y^{\mathrm{S3}} = u_y^{\mathrm{M}} + \frac{L_x}{L_y}\left(u_{x2} - u_{x0}\right) \tag{A.38}$$
$$u_z^{\mathrm{S3}} = u_z^{\mathrm{M}} + \frac{L_x}{L_z}\left(u_{x1} - u_{x0}\right) \tag{A.39}$$

Edge node subset z

$$u_x^{\mathrm{S1}} = u_x^{\mathrm{M}} + u_{x4} - u_{x0} \tag{A.40}$$

$$u_y^{\mathrm{S1}} = u_y^{\mathrm{M}} + \frac{L_x}{L_y}\left(u_{x2} - u_{x0}\right) \tag{A.41}$$

$$u_z^{\mathrm{S1}} = u_z^{\mathrm{M}} + \frac{L_x}{L_z}\left(u_{x1} - u_{x0}\right) \tag{A.42}$$

$$u_x^{\mathrm{S2}} = u_x^{\mathrm{M}} + u_{x2} + u_{x4} - 2u_{x0} \tag{A.43}$$

$$u_y^{\mathrm{S2}} = u_y^{\mathrm{M}} + u_{y2} - u_{y0} + \frac{L_x}{L_y}\left(u_{x2} - u_{x0}\right) \tag{A.44}$$

$$u_z^{\mathrm{S2}} = u_z^{\mathrm{M}} + \frac{L_x}{L_z}\left(u_{x1} - u_{x0}\right) + \frac{L_y}{L_z}\left(u_{y1} - u_{y0}\right) \tag{A.45}$$

$$u_x^{\mathrm{S3}} = u_x^{\mathrm{M}} + u_{x2} - u_{x0} \tag{A.46}$$

$$u_y^{\mathrm{S3}} = u_y^{\mathrm{M}} + u_{y2} - u_{y0} \tag{A.47}$$

$$u_z^{\mathrm{S3}} = u_z^{\mathrm{M}} + \frac{L_y}{L_z}\left(u_{y1} - u_{y0}\right) \tag{A.48}$$

A.2.3 Coupling equation of surface nodes

Analog to the corner nodes and the edge nodes, the remaining nodes are divided into three surface node subsets, depicted in Fig. A.2.2. Each subset contains the master-nodes on the master-surface (green surface in Fig. A.2.2) indicated by the superscript $(\cdot)^{\mathrm{M}}$ and the slave-nodes on the slave-surface (red surface in Fig. A.2.2) indicated by $(\cdot)^{\mathrm{S}}$. The displacements of the slave-nodes are given by equations (A.49) to (A.57).

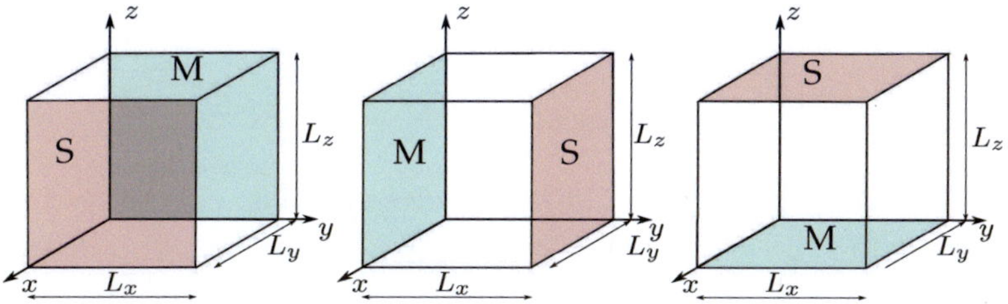

Figure A.2.2: Subsets of the surface nodes.

Surface node subset x

$$u_x^{\mathrm{S}} = u_x^{\mathrm{M}} + u_{x4} - u_{x0} \tag{A.49}$$

$$u_y^{\mathrm{S}} = u_y^{\mathrm{M}} + \frac{L_x}{L_y}\left(u_{x2} - u_{x0}\right) \tag{A.50}$$

$$u_z^{\mathrm{S}} = u_z^{\mathrm{M}} + \frac{L_x}{L_z}\left(u_{x1} - u_{x0}\right) \tag{A.51}$$

Surface node subset y

$$u_x^{\mathrm{S}} = u_x^{\mathrm{M}} + u_{x2} - u_{x0} \tag{A.52}$$

$$u_y^{\mathrm{S}} = u_y^{\mathrm{M}} + u_{y2} - u_{y0} \tag{A.53}$$

$$u_z^{\mathrm{S}} = u_z^{\mathrm{M}} + \frac{L_y}{L_z}\left(u_{y1} - u_{y0}\right) \tag{A.54}$$

Surface node subset z

$$u_x^{\mathrm{S}} = u_x^{\mathrm{M}} + u_{x1} - u_{x0} \tag{A.55}$$

$$u_y^{\mathrm{S}} = u_y^{\mathrm{M}} + u_{y1} - u_{y0} \tag{A.56}$$

$$u_z^{\mathrm{S}} = u_z^{\mathrm{M}} + u_{z1} - u_{z0} \tag{A.57}$$

A.3 Error estimation

The IGFEM approach is validated in Section 8.3 by comparing the numerical simulation result of a laminate for single slip with an analytical solution. The analytical solution is only true in the stationary slip case, thus,

$$\dot{\gamma}(x) = \dot{\gamma}_0 \mathrm{sgn}(\tau) \left\langle \frac{|\tau| - \check{p}}{\tau^{\mathrm{D}}} \right\rangle^m = 0 \tag{A.58}$$

and, therefore, $|\tau| - \check{p} = 0$. The stationary condition cannot be directly transferred to the IGFEM simulation. As a result, an additional stress τ_{err} appears

$$\tau - \check{p} - \underbrace{\left(\frac{\dot{\gamma}(x)}{\dot{\gamma}_0}\right)^{(1/m)} \tau^{\mathrm{D}}}_{:= \tau_{\mathrm{err}}} = \tau - \check{p} - \tau_{\mathrm{err}} = 0, \tag{A.59}$$

where $\tau \geq 0$ is assumed. To keep the effect of the non-zero plastic slip rate as small as possible, material parameters which reduce the additional stress τ_{err} should be chosen. Consequently, it is useful to increase $\dot{\gamma}_0$ and to decrease m as well as τ^{D}. Additionally, the loading rate should be chosen infinitesimal to keep the $\dot{\gamma}(x)$ value small.

A.4 Cahn Hilliard FEM

The weak forms of the equilibrium equations summarized in Box 9.1 read

$$\begin{aligned} r_{\mathrm{c}} = \int_{\mathcal{V}} & w_{\mathrm{c}} \varrho (c - c_n) \\ & + \Delta t \mathrm{grad}(w_{\mathrm{c}}) \cdot (\varrho \boldsymbol{M}_{\mathrm{d}} \mathrm{grad}(\mu_{\mathrm{d}}))\, \mathrm{d}v = 0, \end{aligned} \tag{A.60}$$

$$r_{\check{c}} = \int_{\mathcal{V}} w_{\check{c}} (\check{c} - c) + \lambda_{\mathrm{d}}^2 \mathrm{grad}(w_{\check{c}}) \cdot \mathrm{grad}(\check{c})\, \mathrm{d}v = 0, \tag{A.61}$$

$$r_{\mathrm{u}} = \int_{\mathcal{V}} \boldsymbol{\sigma} \cdot \mathrm{grad}(\boldsymbol{w}_{\mathrm{u}})^{\mathsf{T}} \mathrm{d}v - \int_{\mathcal{A}_{\mathrm{t}}} \boldsymbol{w}_{\mathrm{u}} \cdot \boldsymbol{t} \mathrm{d}a = 0, \tag{A.62}$$

$$r_{\mu} = \int_{\mathcal{V}} w_{\mu} \mu_{\mathrm{e}} - w_{\mu} \frac{1}{2} \boldsymbol{\varepsilon} \cdot \frac{\partial \mathbb{C}}{\partial c} [\boldsymbol{\varepsilon}] \mathrm{d}v = 0 \tag{A.63}$$

and correspond to the residual equations of the Newton's scheme of the global problem. The trial functions w_{c}, $w_{\check{\mathrm{c}}}$, w_{u}, and w_{μ} correspond to the field variables c, $\check{c}$, $\boldsymbol{u}$, and μ_{e}, respectively. Furthermore, the implicite Euler scheme introduced in Section 5.1 was used for the time discretization of $\dot{c}$. The field variables are spatially discretized by shape functions, thus, $c \approx \underline{N}_{\mathrm{c}}\underline{c}$, $\check{c} \approx \underline{N}_{\check{\mathrm{c}}}\underline{\check{c}}$, $\boldsymbol{u} \approx \underline{\underline{N}}_{\mathrm{u}}\underline{u}$, and $\mu_{\mathrm{e}} \approx \underline{N}_{\mu}\underline{\mu}_{\mathrm{e}}$. Here, a vector-matrix notation is adapted, where arrays and matrices are indicated by one and two underlines, respectively. Furthermore, the nodal value of the field variables are denoted by $\underline{c}$, $\underline{\check{c}}$, $\underline{u}$, and $\underline{\mu}_{\mathrm{e}}$. In addition, the gradients of the shape functions are denoted by $\underline{\underline{B}}_{(\cdot)}$. Following the Galerkin FEM, the trial functions are discretized by using the same shape functions as for the corresponding field variables. The spatial discretization of the residual equations of the finite element (e) read

$$\begin{aligned} \underline{r}_{\mathrm{c}}^{(e)} = \int\limits_{\mathcal{V}^{(e)}} & \underline{N}_{\mathrm{c}}^{\mathsf{T}} \underline{N}_{\mathrm{c}} \varrho (\underline{c} - \underline{c}_n) \\ & + \Delta t \underline{\underline{B}}_{\mathrm{c}}^{\mathsf{T}} \varrho \underline{\underline{M}}_{\mathrm{d}} \left(\left(\frac{\partial^2 W_0}{\partial c^2} + H_\chi \right) \underline{\underline{B}}_{\mathrm{c}} \, \underline{c} - H_\chi \underline{\underline{B}}_{\check{\mathrm{c}}} \, \underline{\check{c}} + \underline{\underline{B}}_{\mu} \, \underline{\mu}_{\mathrm{e}} \right) \mathrm{d}v, \end{aligned} \tag{A.64}$$

$$\underline{r}_{\check{\mathrm{c}}}^{(e)} = \int\limits_{\mathcal{V}^{(e)}} \underline{N}_{\check{\mathrm{c}}}^{\mathsf{T}} \underline{N}_{\check{\mathrm{c}}} \underline{\check{c}} - \underline{N}_{\check{\mathrm{c}}}^{\mathsf{T}} \underline{N}_{\mathrm{c}} \underline{c} + \lambda_{\mathrm{d}}^2 \underline{\underline{B}}_{\check{\mathrm{c}}}^{\mathsf{T}} \, \underline{\underline{B}}_{\check{\mathrm{c}}} \underline{\check{c}} \mathrm{d}v, \tag{A.65}$$

$$\underline{r}_{\mathrm{u}}^{(e)} = \int\limits_{\mathcal{V}^{(e)}} \underline{\underline{B}}_{\mathrm{u}}^{\mathsf{T}} \, \underline{\underline{C}}[\underline{\varepsilon}] \mathrm{d}v - \int\limits_{\mathcal{A}_{\mathrm{t}}^{(e)}} \underline{\underline{N}}_{\mathrm{u}} \, \underline{t} \mathrm{d}a, \tag{A.66}$$

$$\underline{r}_{\mu}^{(e)} = \int\limits_{\mathcal{V}^{(e)}} \underline{N}_{\mu}^{\mathsf{T}} \underline{N}_{\mu} \, \underline{\mu}_{\mathrm{e}} - \underline{N}_{\mu}^{\mathsf{T}} \left(\frac{1}{2} \underline{\varepsilon} \frac{\partial \underline{\underline{C}}}{\partial c} [\underline{\varepsilon}] \right) \mathrm{d}v. \tag{A.67}$$

The Newton's scheme requires the algorithmic tangent of the residual equations and reads

$$\begin{pmatrix} \frac{\partial \underline{r}_\mathrm{c}}{\partial \underline{c}} & \frac{\partial \underline{r}_\mathrm{c}}{\partial \underline{\check{c}}} & \frac{\partial \underline{r}_\mathrm{c}}{\partial \underline{u}} & \frac{\partial \underline{r}_\mathrm{c}}{\partial \underline{\mu}_\mathrm{e}} \\ \frac{\partial \underline{r}_\mathrm{\check{c}}}{\partial \underline{c}} & \frac{\partial \underline{r}_\mathrm{\check{c}}}{\partial \underline{\check{c}}} & \frac{\partial \underline{r}_\mathrm{\check{c}}}{\partial \underline{u}} & \frac{\partial \underline{r}_\mathrm{\check{c}}}{\partial \underline{\mu}_\mathrm{e}} \\ \frac{\partial \underline{r}_\mathrm{u}}{\partial \underline{c}} & \frac{\partial \underline{r}_\mathrm{u}}{\partial \underline{\check{c}}} & \frac{\partial \underline{r}_\mathrm{u}}{\partial \underline{u}} & \frac{\partial \underline{r}_\mathrm{u}}{\partial \underline{\mu}_\mathrm{e}} \\ \frac{\partial \underline{r}_\mu}{\partial \underline{c}} & \frac{\partial \underline{r}_\mu}{\partial \underline{\check{c}}} & \frac{\partial \underline{r}_\mu}{\partial \underline{u}} & \frac{\partial \underline{r}_\mu}{\partial \underline{\mu}_\mathrm{e}} \end{pmatrix} \begin{pmatrix} \mathrm{d}\underline{c} \\ \mathrm{d}\underline{\check{c}} \\ \mathrm{d}\underline{u} \\ \mathrm{d}\underline{\mu}_\mathrm{e} \end{pmatrix} = - \begin{pmatrix} \underline{r}_\mathrm{c} \\ \underline{r}_\mathrm{\check{c}} \\ \underline{r}_\mathrm{u} \\ \underline{r}_\mu \end{pmatrix}, \qquad \text{(A.68)}$$

with the entries

$$\begin{aligned} \frac{\partial \underline{r}_c}{\partial \underline{c}} = \int\limits_{\mathcal{V}^{(e)}} \Bigg(& \underline{N}_\mathrm{c}{}^\mathsf{T} \frac{\partial \varrho}{\partial c} \underline{N}_\mathrm{c} (\underline{c} - \underline{c}_n) + \underline{N}_\mathrm{c}{}^\mathsf{T} \varrho \\ & + \Delta t \underline{\underline{B}}_\mathrm{c}^\mathsf{T} \Bigg(\left(\frac{\partial \varrho}{\partial c} \underline{\underline{M}}_\mathrm{d} + \varrho \frac{\partial \underline{\underline{M}}_\mathrm{d}}{\partial c} \right) \Bigg(\left(\frac{\partial^2 W_0}{\partial c^2} + H_\chi \right) \underline{\underline{B}}_\mathrm{c} \underline{c} \\ & - H_\chi \underline{\underline{B}}_\mathrm{\check{c}} \underline{\check{c}} + \underline{\underline{B}}_\mu \underline{\mu}_\mathrm{e} \Bigg) + \varrho \underline{\underline{M}}_\mathrm{d} \left(\frac{\partial^3 W_0}{\partial c^3} \underline{\underline{B}}_\mathrm{c} \underline{c} \right) \Bigg) \Bigg) \underline{N}_\mathrm{c} \\ & + \Delta t \underline{\underline{B}}_\mathrm{c}^\mathsf{T} \underline{\underline{M}}_\mathrm{d} \varrho \left(\frac{\partial^2 W_0}{\partial c^2} + H_\chi \right) \underline{\underline{B}}_\mathrm{c} \mathrm{d}v, \end{aligned} \qquad \text{(A.69)}$$

$$\frac{\partial \underline{r}_c}{\partial \underline{\check{c}}} = \int\limits_{\mathcal{V}^{(e)}} -\Delta t \underline{\underline{B}}_\mathrm{c}{}^\mathsf{T} \varrho \underline{\underline{M}}_\mathrm{d} H_\chi \underline{\underline{B}}_\mathrm{\check{c}} \, \mathrm{d}v, \qquad \text{(A.70)}$$

$$\frac{\partial \underline{r}_c}{\partial \underline{u}} = 0, \qquad \text{(A.71)}$$

$$\frac{\partial \underline{r}_c}{\partial \underline{\mu}_\mathrm{e}} = \int\limits_{\mathcal{V}^{(e)}} -\Delta t \underline{\underline{B}}_\mathrm{c}{}^\mathsf{T} \varrho \underline{\underline{M}}_\mathrm{d} \underline{\underline{B}}_\mu \, \mathrm{d}v, \qquad \text{(A.72)}$$

$$\frac{\partial \underline{r}_{\check{c}}}{\partial \underline{c}} = \int\limits_{\mathcal{V}^{(e)}} -\underline{N}_\mathrm{\check{c}}{}^\mathsf{T} \underline{N}_\mathrm{c} \, \mathrm{d}v, \qquad \text{(A.73)}$$

$$\qquad \text{(A.74)}$$

$$\frac{\partial \underline{r}_{\check{c}}}{\partial \underline{\check{c}}} = \int\limits_{\mathcal{V}^{(e)}} \underline{N}_{\check{c}}{}^{\mathsf{T}} \underline{N}_{\check{c}} + \lambda_{\mathrm{d}}^2 \underline{\underline{B}}_{\check{c}}{}^{\mathsf{T}} \underline{\underline{B}}_{\check{c}} \, \mathrm{d}v, \tag{A.75}$$

$$\frac{\partial \underline{r}_{\check{c}}}{\partial \underline{u}} = 0, \tag{A.76}$$

$$\frac{\partial \underline{r}_{\check{c}}}{\partial \underline{\mu}_{\mathrm{e}}} = 0, \tag{A.77}$$

$$\frac{\partial \underline{r}_{\mathrm{u}}}{\partial \underline{c}} = \int\limits_{\mathcal{V}^{(e)}} \underline{\underline{B}}_{\mathrm{u}}{}^{\mathsf{T}} \left(\frac{\partial \underline{\underline{C}}}{\partial c} [\underline{\varepsilon}] \right) \underline{N}_{\mathrm{c}} \, \mathrm{d}v, \tag{A.78}$$

$$\frac{\partial \underline{r}_{\mathrm{u}}}{\partial \underline{\check{c}}} = 0, \tag{A.79}$$

$$\frac{\partial \underline{r}_{\mathrm{u}}}{\partial \underline{u}} = \int\limits_{\mathcal{V}^{(e)}} \underline{\underline{B}}_{\mathrm{u}}{}^{\mathsf{T}} \frac{\partial \underline{\sigma}}{\partial \underline{\varepsilon}} \underline{\underline{B}}_{\mathrm{u}} \, \mathrm{d}v, \tag{A.80}$$

$$\frac{\partial \underline{r}_{\mathrm{u}}}{\partial \underline{\mu}_{\mathrm{e}}} = 0, \tag{A.81}$$

$$\frac{\partial \underline{r}_{\mu}}{\partial \underline{c}} = \int\limits_{\mathcal{V}^{(e)}} -\underline{N}_{\mu}{}^{\mathsf{T}} \frac{1}{2} \underline{\varepsilon} \frac{\partial^2 \underline{\underline{C}}}{\partial c^2} [\underline{\varepsilon}] \underline{N}_{\mathrm{c}} \, \mathrm{d}v. \tag{A.82}$$

$$\frac{\partial \underline{r}_{\mu}}{\partial \underline{\check{c}}} = 0, \tag{A.83}$$

$$\frac{\partial \underline{r}_{\mu}}{\partial \underline{u}} = \int\limits_{\mathcal{V}^{(e)}} -\underline{N}_{\mu}{}^{\mathsf{T}} \left(\frac{1}{2} \underline{\underline{B}}_{\mathrm{u}}^{\mathsf{S}} \frac{\partial \underline{\underline{C}}}{\partial c} [\underline{\varepsilon}] + \frac{1}{2} \underline{\varepsilon} \frac{\partial \underline{\underline{C}}}{\partial c} \underline{\underline{B}}_{\mathrm{u}}^{\mathsf{S}} \right) \mathrm{d}v, \tag{A.84}$$

$$\frac{\partial \underline{r}_{\mu}}{\partial \underline{\mu}_{\mathrm{e}}} = \int\limits_{\mathcal{V}^{(e)}} \underline{N}_{\mu}{}^{\mathsf{T}} \underline{N}_{\mu} \, \mathrm{d}v, \tag{A.85}$$

with $\underline{\underline{B}}_{\mathrm{u}}^{\mathsf{S}} \underline{u} = \mathrm{sym}(\underline{\underline{B}}_{\mathrm{u}} \underline{u})$.

Frequently used acronyms, symbols, and operators

Acronyms

DS	Directional solidification
FEM	Finite element method
GFEM	Generalized finite element method
GND	Geometrically necessary dislocation
IGFEM	Interface-enriched generalized finite element method
RT	Room temperature
SSD	Statistically stored dislocation
UC	Unit cell
XFEM	Extended finite element method

Latin letters

$\boldsymbol{E}$	Green strain tensor
$\boldsymbol{F}_{\mathrm{e}}, \boldsymbol{F}_{\mathrm{p}}$	Elastic and plastic part of $\boldsymbol{F}$
$\boldsymbol{F}$	Deformation gradient
H_{χ}	Penalty parameter
$\boldsymbol{H}$	Displacement gradient
$J = \det(\boldsymbol{F})$	Determinant of the deformation gradient

$\boldsymbol{M}_{\alpha}$	Schmid tensor of slip system α
$\boldsymbol{M}_{\mathrm{d}}$	Mobility tensor
N_{c}	Number of components in mixture
N_{GS}	Number of slip systems
R_{dis}	Radius of curved dislocation
R	Universal gas constant
$\boldsymbol{S}$	Second Piola-Kirchhoff stress
T	Temperature
W	Free energy density
Q_{C}	Activation energy for creep
$\bar{\boldsymbol{v}}$	Barycentric velocity
$\boldsymbol{b}$	Burgers vector
b	Absolute value of the Burgers vector
$c_{\mathrm{f}}, c_{\mathrm{m}}$	Volume fraction of the fiber and matrix
$\check{c}$	Micromorphic mass concentration
c	Mass concentration
d	Fiber diameter
$\mathbb{C}$	Stiffness tensor
$\boldsymbol{j}$	Mass flux
k_1, k_2	Dislocation storage and annihilation constant
k_{d}	Gradient energy coefficient
m	Stress exponent
$\boldsymbol{n}_{\alpha}, \boldsymbol{d}_{\alpha}$	Slip plane normal and slip direction of system α
$\boldsymbol{n}, \boldsymbol{n}_{\Gamma}$	Normal vector of outer surface and interface
t	Time
$\boldsymbol{v}_{\alpha}$	Velocity of compontent α
$\boldsymbol{x}, \boldsymbol{X}$	Position vectors of a material point in the current and reference placement

Greek letters

$\dot{\bar{\varepsilon}}_{\mathrm{GF}}$	Creep rate where the interface flow rule has been fulfilled for the first time
$\dot{\bar{\varepsilon}}_{\mathrm{min}}$	Minimum creep rate of the composite
$\dot{\varepsilon}_{\mathrm{SS}}$	Steady-state creep rate
$\boldsymbol{\varepsilon}, \boldsymbol{\varepsilon}_{\mathrm{e}}, \boldsymbol{\varepsilon}_{\mathrm{p}}$	Total, elastic, and plastic strain tensor
γ_{acc}	Accumulated plastic slip
$\dot{\gamma}_{\alpha}$	Slip rate of slip system α
λ_{d}	Internal length of Cahn-Hilliard diffusion
μ_{d}	Chemical potential
$\mu_{\mathrm{e}} = \partial W_{\mathrm{e}}/\partial c$	Chemical potential resulting by W_{e}
μ	Shear modulus
$\hat{\phi}_{\beta}$	Nodal degree-of-freedom of $\check{\phi}$
$\phi, \check{\phi}$	Generalized state or internal variables and micromorphic counterpart
π_{G}	Scalar valued generalized micromorphic stress (specified in the corresponding chapters)
ρ_{GND}	Geometrically stored dislocation density
ρ_{α}	Dislocation density of slip system α
ϱ_{α}	Mass density of the component α
ϱ	Total mass density
ρ_{0}	Initial dislocation density
ρ_{ssd}	Statistically stored dislocation density
ρ	Total dislocation density
$\boldsymbol{\sigma}$	Cauchy stress tensor
τ^{F}	Flow stress
τ_{∞}	Theoretical strength
τ_{α}	Schmid stress
$\boldsymbol{\xi}_{\mathrm{G}}$	Vector valued generalized micromorphic stress (specified in the corresponding chapters)

Operators

$\mathrm{Div}\,(\cdot)$	Divergence operator with respect to $\boldsymbol{X}$
$\mathrm{div}\,(\cdot)$	Divergence operator with respect to $\boldsymbol{x}$
$\mathrm{Grad}\,(\cdot)$	Gradient operator with respect to $\boldsymbol{X}$
$\mathrm{grad}\,(\cdot)$	Gradient operator with respect to $\boldsymbol{x}$
$\boldsymbol{A} \cdot \boldsymbol{B}$	Dot product of two tensors $\boldsymbol{A}, \boldsymbol{B}$
$\boldsymbol{A} \otimes \boldsymbol{B}$	Dyadic product of two tensors $\boldsymbol{A}, \boldsymbol{B}$
$\boldsymbol{A} = \mathbb{C}\,[\boldsymbol{B}]$	Linear mapping of a second-order tensor by a fourth-order tensor
$\boldsymbol{AB}$	Composition of two second-order tensors

Bibliography

Abe, F., 2009. Analysis of creep rates of tempered martensitic 9%Cr steel based on microstructure evolution. Mater. Sci. Eng. A 510-511, 64–69.

Aifantis, K. E., Ngan, A. H. W., 2007. Modeling dislocation-grain boundary interactions through gradient plasticity and nanoindentation. Mater. Sci. Eng. A 459, 251–261.

Aifantis, K. E., Soer, W. A., De Hosson, J. T. M., Willis, J. R., 2006. Interfaces within strain gradient plasticity: Theory and experiments. Acta Mater. 54, 5077–5085.

Albiez, J., Erdle, H., Weygand, D., Böhlke, T., 2018. A gradient plasticity creep model accounting for slip transfer/activation at interfaces evaluated for the intermetallic NiAl-9Mo. Manuscript accepted for publication in Int. J. Plast.

Albiez, J., Sprenger, I., Heilmaier, M., Böhlke, T., 2015. One-dimensional simulation of the creep behavior of directionally solidified NiAl-9Mo. Proc. Appl. Math. Mech. 15, 269–270.

Albiez, J., Sprenger, I., Seemüller, C., Weygand, D., Heilmaier, M., Böhlke, T., 2016a. Physically motivated model for creep of directionally solidified eutectics evaluated for the intermetallic NiAl-9Mo. Acta Mat. 110, 377–385.

Albiez, J., Sprenger, I., Weygand, D., Heilmaier, M., Böhlke, T., 2016b. Validation of the applicability of a creep model for directionally solidified eutectics with a lamellar microstructure. Proc. Appl. Math. Mech. 16, 297–298.

Alers, G. A., Neighbours, J. R., Sato, H., 1960. Temperature dependent magnetic contributions to the high field elastic constants of nickel and an Fe-Ni alloy. J. Phys. Chem. Solids 13, 40–55.

Ammar, K., Appolaire, B., Forest, S., Cottura, M., Bouar, Y. L., Finel, A., 2014. Modelling inheritance of plastic deformation during migration of phase boundaries using a phase field method. Meccanica 49, 2699–2717.

Anand, L., Gurtin, M. E., Reddy, B. D., 2015. The stored energy of cold work, thermal annealing, and other thermodynamic issues in single crystal plasticity at small length scales. Int. J. Plast. 64, 1–25.

Appel, F., Brossmann, U., Christoph, U., Eggert, S., Janschek, P., Lorenz, U., Müllauer, J., Oehring, M., Paul, J. D. H., 2000. Recent progress in the development of gamma titanium aluminide alloys. Adv. Eng. Mater. 2, 699–720.

Arzt, E., 1998. Size effects in materials due to microstructural and dimensional constraints: A comparative review. Acta Mater. 46, 5611–5662.

Arzt, E., Dehm, F., Gumbsch, P., Kraft, O., Weiss, D., 2001. Interface controlled plasticity in metals: Dispersion hardening and thin film deformation. Prog. Mater. Sci. 46, 283–307.

Arzt, E., Grahle, P., 1998. High temperature creep behavior of oxide dispersion strengthened NiAl intermetallics. Acta Mater. 46, 2717–2727.

Ashby, M. F., 1970. The deformation of plastically non-homogeneous materials. Phil. Mag. 21, 399–424.

Bachurin, D. V., Weygand, D., Gumbsch, P., 2010. Dislocation-grain boundary interaction in ⟨111⟩ textured thin metal films. Acta Mater. 58, 5232–5241.

Barabash, R. I., Bei, H., Gao, Y., Ice, G. E., George, E. P., 2010a. 3D X-ray microprobe investigation of local dislocation densities and elastic strain gradients in a NiAl-Mo composite and exposed Mo micropillars as a function of prestrain. J. Mater. Res. 25, 199–206.

Barabash, R. I., Bei, H., Gao, Y. F., Ice, G. E., 2010b. Indentation-induced localized deformation and elastic strain partitioning in composites at submicron length scale. Acta Mater. 58, 6784–6789.

Barabash, R. I., Bei, H., Gao, Y. F., Ice, G. E., 2011. Interface strength in NiAl-Mo composites from 3-D X-ray microdiffraction. Scr. Mater. 64, 900–903.

Bartel, T., Menzel, A., Svendsen, B., 2011. Thermodynamic and relaxation-based modeling of the interaction between martensitic phase transformations and plasticity. J. Mech. Phys. Solids 59, 1004–1019.

Bastin, G. F., van Loo, F. J. J., Vrolijk, J. W. G. A., Wolff, L. R., 1978. Crystallography of aligned Fe-Al eutectoid. J. Cryst. Growth 43, 745–751.

Bayerschen, E., Böhlke, T., 2016. Power-law defect energy in a single-crystal gradient plasticity framework: A computational study. Comput. Mech. 58, 13–27.

Bayerschen, E., McBride, A. T., Reddy, B. D., Böhlke, T., 2016. Review on slip transmission criteria in experiments and crystal plasticity models. J. Mater. Sci. 51, 2243–2258.

Bei, H., Barabash, R. I., Ice, G. E., Liu, W., Tischler, J., George, E. P., 2008a. Spatially resolved strain measurements in Mo-alloy micropillars by differential aperture X-ray microscopy. Appl. Phys. Lett. 93, 1–3.

Bei, H., George, E. P., 2005. Microstructures and mechanical properties of a directionally solidified NiAl-Mo eutectic alloy. Acta Mater. 53, 69–77.

Bei, H., George, E. P., 2007. Microstructures and Mechanical Properties of Directionally Solidified Intermetallic Composites. Mater. Sci. Forum 539-543, 1495–1500.

Bei, H., George, E. P., Brown, D. W., Pharr, G. M., Choo, H., Porter, W. D., Bourke, M. A. M., 2005. Thermal-expansion behavior of a directionally solidified NiAl-Mo composite investigated by neutron diffraction and dilatometry. J. Appl. Phys. 97, 1–5.

Bei, H., George, E. P., Pharr, G. M., 2008b. Small-scale mechanical behavior of intermetallics and their composites. Mater. Sci. Eng. A 483-484, 218–222.

Bei, H., Pharr, G. M., George, E. P., 2004. A review of directionally solidified intermetallic composites for high-temperature structural applications. J. Mater. Sci. 39, 3975–3984.

Bei, H., Shim, S., George, E. P., Miller, M., Herbert, E., Pharr, G. M., 2007. Compressive strengths of molybdenum alloy micro-pillars prepared using a new technique. Scr. Mater. 57, 397–400.

Bei, H., Shim, S., Pharr, G. M., George, E. P., 2008c. Effects of pre-strain on the compressive stress-strain response of Mo-alloy single-crystal micropillars. Acta Mater. 56, 4762–4770.

Bei, H., Xia, Y., Barabash, R., Gao, Y., 2016. A tale of two mechanisms: Strain-softening versus strain-hardening in single crystals under small stressed volumes. Scr. Mater. 110, 48–52.

Belytschko, T., Moës, N., Usui, S., Parimi, C., 2001. Arbitrary discontinuities in finite elements. Int. J. Numer. Meth. Eng. 50, 993–1013.

Bertram, A., 1998. An alternative approach to finite plasticity based on material isomorphisms. Int. J. Plast. 52, 353–374.

Bertram, A., 2005. Elasticity and Plasticity of Large Deformations: An Introduction, 1st Edition. Springer, Berlin.

Bertram, A., 2008. Elasticity and Plasticity of Large Deformations: An Introduction, 2nd Edition. Springer, Berlin.

Blum, W., 2008. Mechanisms of Creep Deformation in Steel. In: Creep-Resistant Steels. Cambridge, pp. 365–402.

Brady, M. P., Bei, H., Meisner, R. A., Lance, M. J., Tortorelli, P. F., 2014. Effect of Mo dispersion size and water vapor on oxidation of two-phase directionally solidified NiAl-9Mo in-situ composites. Scr. Mater. 80, 33–36.

Brenner, S. S., 1956. Tensile strength of whiskers. J. Appl. Phys. 27, 1484–1491.

Cahn, J. W., Hilliard, J. E., 1958. Free energy of a nonuniform system. I. Interfacial free energy. J. Chem. Phys. 28, 258–267.

Chen, Y. X., Cui, C. Y., Guo, J. T., Li, D. X., 2004. Microstructure investigation of NiAl-Cr(Mo) interface in a directionally solidified NiAl-Cr(Mo) eutectic alloyed with refractory metal. Mater. Sci. Eng. A 373, 279–285.

Chisholm, C., Bei, H., Lowry, M. B., Oh, J., Syed Asif, S. A., Warren, O. L., Shan, Z. W., George, E. P., Minor, A. M., 2012. Dislocation starvation and exhaustion hardening in Mo alloy nanofibers. Acta Mater. 60, 2258–2264.

Coble, R. L., 1963. A model for boundary diffusion controlled creep in polycrystalline materials. J. Appl. Phys. 34, 1679–1682.

Cottrell, A. H., 1953. Dislocations and Plastic Flow in Crystals. The international series of monographs on physics. Clarendon Pr., Oxford.

Darolia, R., 1994. Sructural Applications of NiAl. J. Mater. Sci. Technol. 10, 157–169.

De Koning, M., Miller, R., Bulatov, V. V., Abraham, F. F., 2002. Modelling grain-boundary resistance in intergranular dislocation slip transmission. Philos. Mag. A 82, 2511–2527.

Dewald, M. P., Curtin, W. A., 2007. Multiscale modelling of dislocation/grain-boundary interactions: I. edge dislocations impinging on Σ11 (1 1 3) tilt boundary in Al. Model. Simul. Mater. Sci. Eng. 15, 193–215.

Dlouhý, A., Eggeler, G., Merk, N., 1995. A micromechanical model for creep in short fiber reinforced aluminium alloys. Acta Metall. Mater. 43, 535–550.

Dudová, M., Barták, T., Kuchařová, K., Dlouhý, A., 2011. Creep in NiAl-Mo eutectic. Proc. 20th Int. Conf. Metall. Mater., Brno, Czech Republic.

Dudová, M., Kuchařová, K., Barták, T., Bei, H., George, E. P., Somsen, C., Dlouhý, A., 2011. Creep in directionally solidified NiAl-Mo eutectics. Scr. Mater. 65, 699–702.

Dunstan, D. J., Bushby, A. J., 2014. Grain size dependence of the strength of metals: The Hall-Petch effect does not scale as the inverse square root of grain size. Int. J. Plast. 53, 56–65.

El-Awady, J. A., 2015. Unravelling the physics of size-dependent dislocation-mediated plasticity. Nat. Commun. 6, 1–9.

Estrin, Y., 1996. 2 – Dislocation-Density-Related Constitutive Modeling. In: Krausz, A., Krausz, K. (Eds.), Unified Constitutive Laws of Plastic Deformation. Academic Press, San Diego, pp. 69–106.

Fish, J., Belytschko, T., 2007. A First Course Finite Elements. John Wiley & Sons, Chichester, West Sussex, England.

Fleck, N. A., Muller, G. M., Ashby, M. F., Hutchinson, J. W., 1994. Strain gradient plasticity: Theory and experiment. Acta Metall. Mater. 42, 475–487.

Forbes, K. R., Glatzel, U., Darolia, R., Nix, W. D., 1996. High-temperature deformation properties of NiAl single crystals. Metall. Mater. Trans. A 27A, 1229–1240.

Forest, S., 2009. Micromorphic approach for gradient elasticity, viscoplasticity, and damage. J. Eng. Mech. 135, 117–131.

Forest, S., 2018. 16 - Use and Abuse of the Method of Virtual Power in Generalized Continuum Mechanics and Thermodynamics. In: Altenbach, H., Pouget, J., Rousseau, M., Collet, B., Michelitsch, T. (Eds.), Generalized Models and Non-Classical Aproaches in Complex Materials 1. Vol. 89. Springer, Cham, Switzerland, pp. 311–334.

Forest, S., Ammar, K., Appolaire, B., 2011. Micromorphic vs. phase-field approaches for gradient viscoplasticity and phase transformations. Lect. Notes Appl. Comput. Mech. 59 LNACM, 69–88.

Fries, T. P., Belytschko, T., 2010. The extended/generalized finite element method: An overview of the method and its applications. Int. J. Numer. Meth. Eng. 84, 253–304.

Frommeyer, G., Rablbauer, R., 2008. High temperature materials based on the intermetallic compound NiAl reinforced by refractory metals for advanced energy conversion technologies. Mater. Technol. 79, 507–513.

Frommeyer, G., Rablbauer, R., Schäfer, H. J., 2010. Elastic properties of B2-ordered NiAl and NiAl-X (Cr, Mo, W) alloys. Intermetallics 18, 299–305.

Frost, H. J., Ashby, M. F., 1982. Deformation-Mechanism Maps: The Plasticity and Creep of Metals and Ceramics, 1st Edition. Pergamon Press, Oxford.

Gao, Y., Bei, H., 2016. Strength statistics of single crystals and metallic glasses under small stressed volumes. Prog. Mater. Sci. 82, 118–150.

Gerlich, D., Fisher, E., 1969. The high temperature elastic moduli of aluminum. J. Phys. Chem. Solids 30, 1197–1205.

Gottschalk, D., McBride, A., Reddy, B., Javili, A., Wriggers, P., Hirschberger, C., 2016. Computational and theoretical aspects of a grain-boundary model that accounts for grain misorientation and grain-boundary orientation. Comp. Mater. Sci. 111, 443–459.

Groot, S. D., Mazur, P., 1984. Non-Equilibrium Thermodynamics. Dover Publications, New York.

Gross, D., Seelig, T., 2011. Bruchmechanik - Mit einer Einführung in die Mikromechanik, 5th Edition. Springer, Berlin Heidelberg.

Guglielmi, P., Ziehmer, M., Lilleodden, E., 2018. On a novel strain indicator based on uncorrelated misorientation angles for correlating dislocation density to local strength. Acta Mater. 150, 195–205.

Gurtin, M. E., 1996. Generalized Ginzburg-Landau and Cahn-Hilliard equations based on a microforce balance. Physica D: Nonlinear Phenomena 92, 178–192.

Gurtin, M. E., 2002. A gradient theory of single-crystal viscoplasticity that accounts for geometrically necessary dislocations. J. Mech. Phys. Solids 50, 5–32.

Gurtin, M. E., 2008. A theory of grain boundaries that accounts automatically for grain misorientation and grain-boundary orientation. J. Mech. Phys. Solids 56, 640–662.

Gurtin, M. E., Anand, L., 2005a. A theory of strain-gradient plasticity for isotropic, plastically irrotational materials. Part I: Small deformations. J. Mech. Phys. Solids 53, 1624–1649.

Gurtin, M. E., Anand, L., 2005b. A theory of strain-gradient plasticity for isotropic, plastically irrotational materials. Part II: Finite deformations. Int. J. Plast. 21, 2297–2318.

Gurtin, M. E., Fried, E., Anand, L., 2010. The Mechanics and Thermodynamics of Continua. Cambridge University Press, New York.

Gurtin, M. E., Needleman, A., 2005. Boundary conditions in small-deformation, single-crystal plasticity that account for the burgers vector. J. Mech. Phys. Solids 53, 1–31.

Gurtin, M. E., Reddy, B. D., 2009. Alternative formulations of isotropic hardening for Mises materials, and associated variational inequalities. Contin. Mech. Thermodyn. 21, 237–250.

Haenschke, T., Gali, A., Heilmaier, M., Krüger, M., Bei, H., George, E. P., 2010. Synthesis and characterization of lamellar and fibre-reinforced NiAl-Mo and NiAl-Cr. J. Phys. Conf. Ser. 240, 1–4.

Hairer, E., Nørsett, S. P., Wanner, G., 1993. Solving Ordinary Differential Equations I. Springer, Berlin.

Hall, E. O., 1951. The deformation and ageing of mild steel: III discussion of results. Proc. Phys. Soc. B 64, 747–753.

Herring, C., 1950. Diffusional viscosity of a polycrystalline solid. J. Appl. Phys. 21, 437–445.

Hill, R., 1963. Elastic properties of reinforced solids: Some theoretical principles. J. Mech Phys. Solids 11, 357–372.

Hirth, J. P., Lothe, J., 1992. Theory of Dislocations, 2nd Edition. Krieger, Florida.

Hu, L., Hu, W., Gottstein, G., Bogner, S., Hollad, S., Bührig-Polaczek, A., 2012. Investigation into microstructure and mechanical properties of NiAl-Mo composites produced by directional solidification. Mater. Sci. Eng. A 539, 211–222.

Hu, L., Zhang, G., Hu, W., Gottstein, G., Bogner, S., Bührig-Polaczek, A., 2013. Tensile creep of directionally solidified NiAl-9Mo in situ composites. Acta Mater. 61, 7155–7165.

Hull, D., Bacon, D. J., 2011. Introduction to Dislocations. Butterworth-Heinemann, Oxford.

Hutter, K., Jöhnk, K., 2004. Continuum Methods of Physical Modeling: Continuum Mechanics, Dimensional Analysis, Turbulence. Springer, Berlin Heidelberg.

Jackson, K. A., Hunt, J. D., 1966. Lamellar and Rod Eutectic Growth. Trans. Metall. Soc. AIME 236, 1129–1142.

Johanns, K. E., Sedlmayr, A., Sudharshan Phani, P., Mönig, R., Kraft, O., George, E. P., Pharr, G. M., 2011. In-situ tensile testing of single-crystal molybdenum-alloy fibers with various dislocation densities in a scanning electron microscope. J. Mater. Res. 27, 508–520.

Johnson, D. R., Chen, X. F., Oliver, B. F., Noebe, R. D., Whittenberger, J. D., 1995. Processing and mechanical properties of in-situ composites from the NiAlCr and the NiAl(Cr,Mo) eutectic systems. Intermetallics 3, 99–113.

Johnson, L., Ashby, M. F., 1968. The stress at which dislocations multiply in well-annealed metal crystals. Acta Metall. 16, 219–225.

Joslin, S., Chen, X., Oliver, B., Noebe, R., 1995. Fracture behavior of directionally solidified NiAl-Mo and NiAl-V eutectics. Mater. Sci. Eng. A 196, 9–18.

Kaessmair, S., Steinmann, P., 2016. Comparative computational analysis of the Cahn–Hilliard equation with emphasis on C^1-continuous methods. J. Comput. Phys. 322, 783–803.

Kocks, U. F., Mecking, H., 2003. Physics and phenomenology of strain hardening: The FCC case. Prog. Mater. Sci. 48, 171–273.

Kraft, O., Gruber, P. A., Mönig, R., Weygand, D., 2010. Plasticity in confined dimensions. Annu. Rev. Mater. Res. 40, 293–317.

Krawietz, A., 1986. Materialtheorie - Mathematische Beschreibung des phänomenologischen thermomechanischen Verhaltens, 1st Edition. Springer, Berlin.

Kwon, J., Bowers, M. L., Brandes, M. C., McCreary, V., Robertson, I. M., Sudharshan Phani, P., Bei, H., Gao, Y. F., Pharr, G. M., George, E. P., Mills, M. J., 2015. Characterization of dislocation structures and deformation mechanisms in as-grown and deformed directionally solidified NiAl-Mo composites. Acta Mater. 89, 315–326.

Li, X., Bottler, F., Spatschek, R., Schmitt, a., Heilmaier, M., Stein, F., 2017. Coarsening kinetics of lamellar microstructures: Experiments and simulations on a fully-lamellar Fe-Al in situ composite. Acta Mater. 127, 230–243.

Lilleodden, E. T., Nix, W. D., 2006. Microstructural length-scale effects in the nanoindentation behavior of thin gold films. Acta Mater. 54, 1583–1593.

Loya, P. E., Xia, Y. Z., Peng, C., Bei, H., Zhang, P., Zhang, J., George, E. P., Gao, Y. F., Lou, J., 2014. Yield strength dependence on strain rate of molybdenum-alloy nanofibers. Appl. Phys. Lett. 104, 1–5.

Malyar, N., Dehm, G., Kirchlechner, C., 2017. Strain rate dependence of the slip transfer through a penetrable high angle grain boundary in copper. Scr. Mater. 138, 88–91.

McBride, A. T., Gottschalk, D., Reddy, B. D., Wriggers, P., Javili, A., 2016. Computational and theoretical aspects of a grain-boundary model at finite deformations. Tech. Mech. 36, 102–119.

Mecking, H., Kocks, U. F., 1981. Kinetics of flow and strain-hardening. Acta Metall. 29, 1865–1875.

Mecking, H., Kocks, U. F., Fischer, H., 1976. Hardening, recovery, and creep in fcc mono- and polycrystals. Int. Conf. strength Met. Alloy. 8, 334–339.

Miracle, D. B., 1993. Overview No. 104 The physical and mechanical properties of NiAl. Acta Metall. Mater. 41, 649–684.

Misra, A., Gibala, R., 2000. Plasticity in multiphase intermetallics. Intermetallics 8, 1025–1034.

Misra, A., Wu, Z. L., Kush, M. T., Gibala, R., 1997. Microstructures and mechanical properties of directionally solidified NiAl-Mo and NiAl-Mo(Re) eutectic alloys. Mater. Sci. Eng. A 239-240, 75–87.

Misra, A., Wu, Z. L., Kush, M. T., Gibala, R., 1998. Deformation and fracture behaviour of directionally solidified NiAl-Mo and NiAl-Mo(Re) eutectic composites. Philos. Mag. A 78, 533–550.

Moës, N., Dolbow, J., Belytschko, T., 1999. A finite element method for crack growth without remeshing. Int. J. Numer. Meth. Eng. 46, 131–150.

Mukherjee, A. K., 2002. An examination of the constitutive equation for elevated temperature plasticity. Mater. Sci. Eng. A 322, 1–22.

Müller, I., 1967. On the Entropy Inequality. Archive for Rational Mechanics and Analysis 26, 118–141.

Nabarro, F. R. N., 1948. Deformation of Crystals by the Motion of Single Ions. In: Report of a Conference on the Strength of Solids. The Physical Society, London.

Nabarro, F. R. N., de Villiers, H. L., 1995. The Physics of Creep. Taylor & Francis, London.

Naumenko, K., Altenbach, H., 2016. Modeling High Temperature Materials Behavior for Structural Analysis. Springer, Switzerland.

Noebe, R., Bowman, R., Nathal, M., 1993. Physical and mechanical properties of the B2 compound NiAl. Int. Mater. Rev. 38, 193–232.

Nye, J., 1953. Some geometrical relations in dislocated crystals. Acta Metall. 1, 153–162.

Ohno, N., Okumura, D., 2007. Higher-order stress and grain size effects due to self-energy of geometrically necessary dislocations. J. Mech. Phys. Solids 55, 1879–1898.

Orowan, E., 1934a. Zur Kristallplastizität. I. Zeitschrift für Physik 89, 605–613.

Orowan, E., 1934b. Zur Kristallplastizität. II. Zeitschrift für Physik 89, 614–633.

Orowan, E., 1934c. Zur Kristallplastizität. III. Zeitschrift für Physik 89, 634–659.

Ortiz, M., Repetto, E. A., 1999. Nonconvex energy minimization and dislocation structures in ductile single crystals. J. Mech. Phys. Solids 47, 397–462.

Ostoja-Starzewski, M., 2002. Microstructural randomness versus representative volume element in thermomechanics. J. Appl. Mech. 69, 25–35.

Parthasarathy, T. A., Rao, S. I., Dimiduk, D. M., Uchic, M. D., Trinkle, D. R., 2007. Contribution to size effect of yield strength from the stochastics of dislocation source lengths in finite samples. Scr. Mater. 56, 313–316.

Peerlings, R. H. J., De Borst, R., Brekelmans, W. A. M., De Vree, J. H. P., 1996. Gradient enhanced damage for quasi-brittle materials. Int. J. Numer. Methods Eng. 39, 3391–3403.

Petch, N. J., 1953. The cleavage strength of polycrystals. J. Iron Steel Institute 174, 25–28.

Pugh, J. W., 1955. The tensile properties of molybdenum at elevated temperatures. Trans. ASM 47, 984–1001.

Raj, S. V., 2002. Power-law and exponential creep in class M materials: Discrepancies in experimental observations and implications for creep modeling. Mater. Sci. Eng. A 322, 132–147.

Reddy, B. D., 2011a. The role of dissipation and defect energy in variational formulations of problems in strain-gradient plasticity. Part 1: Polycrystalline plasticity. Contin. Mech. Thermodyn. 23, 527–549.

Reddy, B. D., 2011b. The role of dissipation and defect energy in variational formulations of problems in strain-gradient plasticity. Part 2: Single-crystal plasticity. Contin. Mech. Thermodyn. 23, 551–572.

Reddy, B. D., Ebobisse, F., McBride, A., 2008. Well-posedness of a model of strain gradient plasticity for plastically irrotational materials. Int. J. of Plast. 24, 55–73.

Rinaldi, A., 2011. Effects of dislocation density and sample-size on plastic yielding at the nanoscale: A Weibull-like framework. Nanoscale 3, 4817–4823.

Rusović, N., Warlimont, H., 1977. The Elastic Behaviour of β_2-NiAl Alloys. Phys. Stat. Sol. (a) 44, 609–619.

Scherf, A., Janda, D., Baghaie Yazdi, M., Li, X., Stein, F., Heilmaier, M., 2015. Oxidation behavior of binary aluminium-rich Fe–Al alloys with a fine-scaled, lamellar microstructure. Oxid. Met. 83, 559–574.

Scherf, A., Kauffmann, A., Kauffmann-Weiss, S., Scherer, T., Li, X., Stein, F., Heilmaier, M., 2016. Orientation relationship of eutectoid FeAl and $FeAl_2$. J. Appl. Crystallogr. 49, 442–449.

Schmitt, A., Kumar, K. S., Kauffmann, A., Li, X., Stein, F., Heilmaier, M., 2017. Creep of binary Fe-Al alloys with ultrafine lamellar microstructures. Intermetallics 90, 180–187.

Seemüller, C., Heilmaier, M., Haenschke, T., Bei, H., Dlouhý, A., George, E. P., 2013. Influence of fiber alignment on creep in directionally solidified NiAl-10Mo in-situ composites. Intermetallics 35, 110–115.

Shang, Z., Shen, J., Zhang, J., Wang, L., Wang, L., Fu, H., 2014. Effect of microstructures on the room temperature fracture toughness of NiAl–32Cr–6Mo hypereutectic alloy directionally solidified at different withdrawal rates. Mater. Sci. Eng. A 611, 306–312.

Simo, J., Hughes, T., 1998. Computational Inelasticity. Vol. 7. Springer, New York.

Soghrati, S., Aragón, A. M., Armando Duarte, C., Geubelle, P. H., 2012. An interface-enriched generalized FEM for problems with discontinuous gradient fields. Int. J. Numer. Mech. Eng. 89, 991–1008.

Strouboulis, T., Babuška, I., Copps, K., 2000. The design and analysis of the generalized finite element method. Comput. Meth. Appl. Mech. Eng. 181, 43–69.

Sudharshan Phani, P., Johanns, K. E., Duscher, G., Gali, A., George, E. P., Pharr, G. M., 2011. Scanning transmission electron microscope observations of defects in as-grown and pre-strained Mo alloy fibers. Acta Mater. 59, 2172–2179.

Sudharshan Phani, P., Johanns, K. E., George, E. P., Pharr, G. M., 2013. A simple stochastic model for yielding in specimens with limited number of dislocations. Acta Mater. 61, 2489–2499.

Svendsen, B., Bargmann, S., 2010. On the continuum thermodynamic rate variational formulation of models for extended crystal plasticity at large deformation. J. Mech. Phys. Solids 58, 1253–1271.

Taylor, G. I., 1934. The mechanism of plastic deformation of crystals. Part I.-Theoretical. Proc. R. Soc. A Math. Phys. Eng. Sci. 145, 362–387.

Tschopp, M. A., Tucker, G. J., McDowell, D. L., 2008. Atomistic simulations of tension-compression asymmetry in dislocation nucleation for copper grain boundaries. Comput. Mater. Sci. 44, 351–362.

Ubachs, R. L. J. M., Schreurs, P. J. G., Geers, M. G. D., 2004. A nonlocal diffuse interface model for microstructure evolution of tin-lead solder. J. Mech. Phys. Solids 52, 1763–1792.

Ubachs, R. L. J. M., Schreurs, P. J. G., Geers, M. G. D., 2005. Phase field dependent viscoplastic behaviour of solder alloys. Int. J. Solids Struct. 42, 2533–2558.

von Blanckenhagen, B., Gumbsch, P., Arzt, E., 2001. Dislocation sources in discrete dislocation simulations of thin-film plasticity and the Hall-Petch relation. Model. Simul. Mater. Sci. Eng. 9, 157–169.

von Blanckenhagen, B., Gumbsch, P., Arzt, E., 2003. Dislocation sources and the flow stress of polycrystalline thin metal films. Philos. Mag. Lett. 83, 1–8.

Voyiadjis, G. Z., Faghihi, D., Zhang, Y., 2014. A theory for grain boundaries with strain-gradient plasticity. Int. J. Solids Struct. 51, 1872–1889.

Waffenschmidt, T., Polindara, C., Menzel, A., Blanco, S., 2014. A gradient-enhanced large-deformation continuum damage model for fibre-reinforced materials. Comput. Meth. Appl. Mech. Eng. 268, 801–842.

Walter, J. L., Cline, H. E., 1970. The effect of solidification rate on structure and high-temperature strength of the eutectic NiAl-Cr. Metall. Mater. Trans. 1, 1221–1229.

Wang, L., Shen, J., Zhang, G., Zhang, Y., Guo, L., Ge, Y., Gao, L., Fu, H., 2018. Stability of lamellar structure of directionally solidified NiAl-28Cr-6Mo eutectic alloy at different withdrawal rates and temperatures. Intermetallics 94, 83–91.

Wen, C. E., Yasue, K., Lin, J. G., Zhang, Y. G., Chen, C. Q., 2000. The effect of lamellar spacing on the creep behavior of a fully lamellar TiAl alloy. Intermetallics 8, 525–529.

Wern, H., 2004. Single Crystal Elastic Constants and Calculated Bulk Properties: A Handbook. Logos-Verl., Berlin.

Whittenberger, J., Raj, S., Locci, I., Salem, J., 1999. Effect of growth rate on elevated temperature plastic flow and room temperature fracture toughness of directionally solidified NiAl-31Cr-3Mo. Intermetallics 7, 1159–1168.

Whittenberger, J. D., 1987. Effect of composition and grain size on slow plastic flow properties of NiAl between 1200 and 1400 K. J. Mater. Sci. 22, 394–402.

Whittenberger, J. D., 1988. The influence of grain size and composition on 1000 to 1400 K slow plastic flow properties of NiAl. J. Mater. Sci. 23, 235–240.

Whittenberger, J. D., Noebe, R. D., Cullers, C. L., Kumar, K. S., Mannan, S. K., 1991. 1000 to 1200 K time-dependent compressive deformation of single-crystalline and polycrystalline B2 Ni-40Al. Metall. Trans. A 22, 1595–1607.

Wulfinghoff, S., Bayerschen, E., Böhlke, T., 2013. A gradient plasticity grain boundary yield theory. Int. J. Plast. 51, 33–46.

Wulfinghoff, S., Böhlke, T., 2012. Equivalent plastic strain gradient enhancement of single crystal plasticity: Theory and numerics. Proc. R. Soc. A Math. Phys. Eng. Sci. 468, 2682–2703.

Wulfinghoff, S., Böhlke, T., 2015. Gradient crystal plasticity including dislocation-based work-hardening and dislocation transport. Int. J. Plast. 69, 152–169.

Wulfinghoff, S., Forest, S., Böhlke, T., 2015. Strain gradient plasticity modeling of the cyclic behavior of laminate microstructures. J. Mech. Phys. Solids 79, 1–20.

Xia, Z., Zhang, Y., Ellyin, F., 2003. A unified periodical boundary conditions for representative volume elements of composites and applications. Int. J. Solids Struct. 40, 1907–1921.

Xu, J., Zeng, W., Sun, X., Jia, Z., Zhou, J., 2015. Static coarsening behavior of the lamellar alpha in Ti-17 alloy. J. Alloys Compd. 631, 248–254.

Yang, J. M., Jeng, S. M., Bain, K., Amato, R. A., 1997. Microstructure and mechanical behavior of in-situ directional solidified NiAl/Cr(Mo) eutectic composite. Acta Mater. 45, 295–308.

Yu, D., An, K., Chen, X., Bei, H., 2016. Phase-specific deformation behavior of a NiAl–Cr(Mo) lamellar composite under thermal and mechanical loads. J. Alloys Comp. 656, 481–490.

Yu, D., Bei, H., Chen, Y., George, E. P., An, K., 2014. Phase-specific deformation behavior of a relatively tough NiAl-Cr(Mo) lamellar composite. Scr. Mater. 84-85, 59–62.

Zhang, J. F., Shen, J., Shang, Z., Feng, Z. R., Wang, L. S., Fu, H. Z., 2012. Microstructure and room temperature fracture toughness of directionally solidified NiAl-Mo eutectic in situ composites. Intermetallics 21, 18–25.

Zhang, J. F., Shen, J., Shang, Z., Wang, L., Fu, H. Z., 2013. Directional solidification and characterization of NiAl-9Mo eutectic alloy. Trans. Nonferrous Met. Soc. China (English Ed.) 23, 3499–3507.

Zhang, J. F., Shen, J., Shang, Z., Wang, L., Fu, H. Z., 2015. Elevated temperature tensile properties and deformation of directionally solidified NiAl–Mo in-situ composites. Mater. Charact. 99, 160–165.

Zhu, T. T., Bushby, A. J., Dunstan, D. J., 2008. Materials mechanical size effects: A review. Mater. Technol. 23, 193–209.

Schriftenreihe Kontinuumsmechanik im Maschinenbau
Karlsruher Institut für Technologie (KIT)
(ISSN 2192-693X)

(ISSN 2192-693X)

Herausgeber: Prof. Dr.-Ing. Thomas Böhlke

Band 1 Felix Fritzen
Microstructural modeling and computational homogenization of the physically linear and nonlinear constitutive behavior of micro-heterogeneous materials. 2011
ISBN 978-3-86644-699-1

Band 2 Rumena Tsotsova
Texturbasierte Modellierung anisotroper Fließpotentiale. 2012
ISBN 978-3-86644-764-6

Band 3 Johannes Wippler
Micromechanical finite element simulations of crack propagation in silicon nitride. 2012
ISBN 978-3-86644-818-6

Band 4 Katja Jöchen
Homogenization of the linear and non-linear mechanical behavior of polycrystals. 2013
ISBN 978-3-86644-971-8

Band 5 Stephan Wulfinghoff
Numerically Efficient Gradient Crystal Plasticity with a Grain Boundary Yield Criterion and Dislocation-based Work-Hardening. 2014
ISBN 978-3-7315-0245-6

Band 6 Viktor Müller
Micromechanical modeling of short-fiber reinforced composites. 2016
ISBN 978-3-7315-0454-2

Die Bände sind unter www.ksp.kit.edu als PDF frei verfügbar oder als Druckausgabe bestellbar.

Band 7	Florian Rieger **Work-hardening of dual-phase steel.** 2016 ISBN 978-3-7315-0513-6
Band 8	Vedran Glavas **Micromechanical Modeling and Simulation of Forming Processes.** 2017 ISBN 978-3-7315-0602-7
Band 9	Eric Bayerschen **Single-crystal gradient plasticity with an accumulated plastic slip: Theory and applications.** 2017 ISBN 978-3-7315-0606-5
Band 10	Bartholomäus Brylka **Charakterisierung und Modellierung der Steifigkeit von langfaserverstärktem Polypropylen.** 2017 ISBN 978-3-7315-0680-5
Band 11	Rudolf Neumann **Two-Scale Thermomechanical Simulation of Hot Stamping.** 2017 ISBN 978-3-7315-0714-7
Band 12	Mauricio Lobos Fernández **Homogenization and materials design of mechanical properties of textured materials based on zeroth-, first- and second-order bounds of linear behavior.** 2018 ISBN 978-3-7315-0770-3
Band 13	Malte Schemmann **Biaxial Characterization and Mean-field Based Damage Modeling of Sheet Molding Compound Composites.** 2018 ISBN 978-3-7315-0818-2
Band 14	Jürgen Albiez **Finite element simulation of dislocation based plasticity and diffusion in multiphase materials at high temperature.** 2019 ISBN 978-3-7315-0918-9